Heat Transfer and Fluid Dynamics in Boiling Systems

Heat Transfer and Fluid Dynamics in Boiling Systems

Editors

Vladimir Serdyukov
Fedor Ronshin

MDPI

Basel • Beijing • Wuhan • Barcelona • Belgrade • Novi Sad • Cluj • Manchester

Editors
Vladimir Serdyukov
Kutateladze Institute of Thermophysics
Novosibirsk
Russia

Fedor Ronshin
Kutateladze Institute of Thermophysics
Novosibirsk
Russia

Editorial Office
MDPI
St. Alban-Anlage 66
4052 Basel, Switzerland

This is a reprint of articles from the Special Issue published online in the open access journal *Energies* (ISSN 1996-1073) (available at: https://www.mdpi.com/journal/energies/special_issues/HT_FD_BS).

For citation purposes, cite each article independently as indicated on the article page online and as indicated below:

Lastname, A.A.; Lastname, B.B. Article Title. *Journal Name* **Year**, *Volume Number*, Page Range.

ISBN 978-3-7258-0717-8 (Hbk)
ISBN 978-3-7258-0718-5 (PDF)
doi.org/10.3390/books978-3-7258-0718-5

Contents

Preface

Boiling is a highly efficient heat transfer mechanism, which has wide-ranging applications in diverse industrial and technological sectors such as thermal desalination, heat and nuclear energies, heat pipes, and heat exchangers. It is particularly acknowledged as the most effective method for the thermal management of high-heat-flux devices like high-power electronics. Ongoing research endeavors include the investigation of pool and flow boiling mechanisms and the development of techniques to enhance boiling performance. The process of boiling is intricate and influenced by various factors, with surface topology and wettability playing a particularly critical role. Accordingly, the modification of surfaces for boiling applications constitutes a prominent area of research.

This Special Issue presents a collection of carefully selected papers that delve into heat transfer and fluid dynamics during boiling phenomena. The Special Issue comprises state-of-the-art experimental and numerical research contributions, with a specific emphasis on recent advancements in enhancing heat transfer during boiling under different conditions, especially through novel approaches to surface modification for heating purposes. The Guest Editors extend their gratitude to the authors for submitting high-quality research articles and to the referees for providing valuable reviews that facilitated the selection process. It is anticipated that this Special Issue will prove invaluable to the community and inspire further progress in the rapidly evolving field of boiling. Last but not least, we want to thank Mr. Chester Zheng—the Assistant Editor, MDPI—for his dedication to this Special Issue. He has been a major supporter of this Special Issue, and we are indebted to him.

One of the ways to enhance nucleate boiling performance and increase critical heat fluxes is using wettability patterned surfaces or coatings, so-called biphilic surfaces. In the research of Chinnov et al., flat surfaces with different patterns of hydrophobic spots were employed for the experimental investigation of water boiling heat transfer. The results show that boiling heat transfer on the test biphilic surfaces was significantly higher (up to 600%) than on base surfaces. The highest heat transfer efficiency was detected for the surface with the largest number of hydrophobic spots. After long-term experiments (up to 3 years), the heat transfer coefficient on the obtained surfaces remained higher than on the smooth copper surface. Biphilic surfaces with arrays of cavities formed due to laser ablation turned out to be the most stable during prolonged contact with boiling water.

To study the processes of boiling on a smooth surface with contrast wettability, Fedoseev et al. developed a hybrid model based on the Lattice Boltzmann method and heat transfer equation. To find the optimal configuration of the biphilic surface, at the first stage, a numerical simulation was carried out for a single lyophobic zone on a lyophilic surface. The dependences of the bubble departure frequency and the departure diameter of the bubble on the width of the lyophobic zone were obtained, and its optimal size was determined. At the next stage, the boiling process on an extended surface was studied in the presence of several lyophobic zones of a given size with different distances between them. It was shown that, in the region of moderate surface superheat, the intensity of heat transfer on biphilic surfaces can be several times (more than four times) higher compared to surfaces with homogeneous wettability. Based on numerical calculations, an optimal configuration of the biphilic surface with the ratios of the lyophobic zones' width of the order of 0.16 and the distance between the lyophobic zones in the range of 0.9–1.3 to the bubble departure diameter was found.

In the research of Serdyukov et al., the results of an experimental study on the evolution of a nonstationary temperature field during ethanol pool boiling in a pressure range of 12–101.2 kPa are presented. Experimental data were obtained using infrared thermography with high temporal and spatial resolutions, which made it possible to reconstruct the distribution of the heat flux density and to study the influence of pressure reduction on the local heat transfer rate in the vicinity of the triple contact line under vapor bubbles for the first time. It was shown that, for all studied pressures, a significant heat flux density is removed from the heating surface due to microlayer evaporation, which exceeds the input heat power by a factor of 3.3–27.7, depending on the pressure. Meanwhile, the heat transfer rate in the area of the microlayer evaporation significantly decreases with the pressure reduction. Estimates of the microlayer profile based on the heat conduction equation were made, which showed the significant increase in the microlayer thickness with the pressure reduction.

Nunes et al. analyzed the influence of different heights of square micro pin fins with an aligned array and investigates their influence on pressure drop and heat transfer behavior. The HFE-7100 was used as the working fluid, and the pressure drop and surface temperature behavior are analyzed for different mass fluxes and inlet subcooling. There is good agreement between the experimental results and the numerical analysis, with a mean absolute error of 6% for all the considered parameters. For the two-phase flow condition, experimental tests were performed, and for the highest subcooling, an increase in mass flux causes an enhancement in the heat transfer for low heat flux; by increasing heat flux, there is a gradual predominance of boiling heat transfer over convection as the heat transfer mechanism. The pressure drop drastically increases with the vapor amount flowing into the system, regardless of the pin fin height; the boiling curves for the higher fin height show a much smaller slope and a smaller wall superheat than the fin with the smallest height, and consequently, a high heat transfer performance.

In the research of Vostretsov et al., the influence of parameters such as the total flow rate of phases, the ratio of flow rates, and residence time on mass transfer during the two-phase flow of immiscible liquids in a T-shaped microchannel were investigated using the micro-LIF technique. The study focused on the plug flow regime, where a 70% water–glycerol solution was used as the dispersed phase, and tri-n-butyl phosphate (TBP) was used as the carrier phase. Using the obtained data, the extraction efficiency and overall volumetric mass transfer coefficient and established dependencies were determined demonstrating the effect of the flow-rate ratio, total flow rate, and the residence time on the mass transfer rate and extraction efficiency. Finally, a model for the overall volumetric mass transfer coefficient corresponding to the set of liquids used with an R-squared value of 0.966 was proposed.

Volodin et al. presented the experimental results of combining a basic microstructure with partly closed pores and a mesh covering for heat transfer enhancement at the film flow of a refrigerant mixture. All experimental series were carried out using a binary mixture of R114 and R21 refrigerants. It was shown that a microstructured surface with a fin pitch of 200 μm, fin height of 220 μm, and longitudinal knurling pitch of 160 μm, created using deformational cutting, demonstrates significant heat transfer enhancement: up to four times compared to a smooth surface. However, adding a mesh covering with an aperture of 220 μm and a wire diameter of 100 μm reduces the intensification. The mesh covering overlaid on a smooth surface also does not provide heat transfer enhancement as compared to the smooth surface itself. The heat transfer coefficient values obtained for basic microstructured surfaces were compared with the dependencies available in the literature for predicting pool boiling heat transfer on microfinned surfaces.

In the research of Misyura et al., the results of the experimental study of the effect of the droplet evaporation rate on wall temperature in the range 20–90 °C and of the concentration of surfactant in an aqueous solution of sodium lauryl sulfate (SLS) from 0 to 10,000 ppm are presented. It is shown for the first time that an inversion of the evaporation rate related to the droplet diameter occurs with increasing wall temperature. The influence of key factors on the evaporation of a water droplet with SLS changes with temperature. Thus, at a slightly heated wall, the growth of the droplet diameter becomes predominant. At high heat flux, the role of nonisothermality is predominant. To determine the individual influence of the surfactant on the partial pressure of water vapor, experiments on the evaporation of a liquid layer were carried out.

The effect of coal hydrophilic particles in water-glycerol drops on the maximum diameter of spreading along a hydrophobic solid surface is experimentally studied by Ashikhmin et al. via analyzing the velocity of internal flows using Particle Image Velocimetry (PIV). The impact of particle-laden drops on a solid surface occurred at Weber numbers (We) from 30 to 120. It revealed the interrelated influence of We and the concentration of coal particles on changes in the maximum absolute velocity of internal flows in a drop within the kinetic and spreading phases of the drop–wall impact. The kinetic energy of the translational motion of coal particles in a spreading drop compensates for the energy expended by the drop in sliding friction along the wall. An increase in We contributes to more noticeable differences in the convection velocities in spreading drops. When the drop spreading diameter rises at the maximum velocity of internal flows, a growth of the maximum spreading diameter occurs. The presence of coal particles causes a general tendency to reduce drop spreading.

Kotov et al. studied the effect of temperature and pressure on the nucleation of the vapor phase and the velocity of the vapor front in the initial stage of activated boiling-up of n-pentane on the surface of a quartz fiber. Using a developed approach combining the "pump-probe" and laser Doppler velocimetry methods, this velocity was tracked in the course of sequential change in the degree of superheating with respect to the liquid–vapor equilibrium line. The studied interval according to the degree of superheating was 40–100 °C (at atmospheric pressure). An increase in temperature at a given pressure was found to lead to an increase in the speed of the transition process with a coefficient of about 0.2 m/s per degree, while an increase in pressure at a given temperature leads to a decrease in the transition process speed with a coefficient of 25.8 m/s per megapascal. The advancement of the vapor front velocity measurements to sub-microsecond intervals from the first signs of boiling-up did not confirm the existence of a Rayleigh expansion stage with a constant velocity.

The effect of positive (adverse) and negative (favorable) longitudinal pressure gradients on the structure and heat transfer of gas-droplet (air and water) flow in an axisymmetric duct with sudden expansion were examined by Pakhomov and Terekhov. The superimposed pressure gradient has a large influence on the flow structure and heat transfer in a two-phase mist flow in both a confuser and a diffuser. A narrowing of the confuser angle leads to significant suppression of flow turbulence (more than four times that of the gas-drop flow after sudden pipe expansion without a pressure gradient at $\varphi = 0°$). Recirculation zone length decreases significantly compared to the gas-droplet flow without a longitudinal pressure gradient (by up to 30%), and the locus of the heat-transfer maximum shifts slightly downstream, and roughly aligns with the reattachment point of the two-phase flow. Growth of the diffuser opening angle leads to additional production of kinetic energy of gas flow turbulence (almost twice as much as gas-droplet flow after a sudden pipe expansion at $\varphi = 0°$). The length of the flow recirculating region in the diffuser increases significantly compared to the separated gas-droplet flow without a pressure gradient ($\varphi = 0°$), and the location of maximum heat transfer shifts downstream in the diffuser.

Vladimir Serdyukov and Fedor Ronshin
Editors

energies

MDPI

Article

Boiling Heat Transfer Enhancement on Biphilic Surfaces

Evgeny A. Chinnov [1], Sergey Ya. Khmel [1], Victor Yu. Vladimirov [1,*], Aleksey I. Safonov [1], Vitaliy V. Semionov [1], Kirill A. Emelyanenko [2], Alexandre M. Emelyanenko [2] and Ludmila B. Boinovich [2]

1 Kutateladze Institute of Thermophysics, Siberian Branch of the Russian Academy of Sciences, 1 Lavrentyev Ave., 630090 Novosibirsk, Russia
2 A. N. Frumkin Institute of Physical Chemistry and Electrochemistry, Russian Academy of Sciences, Leninsky Prospect 31 Bldg. 4, 119071 Moscow, Russia
* Correspondence: victor.lipps@gmail.com; Tel.: +7-9137723236

Abstract: Flat surfaces with different patterns of hydrophobic spots were employed for experimental investigation of boiling heat transfer. In one case, hydrophobic spots were created on a smooth copper surface and on a surface coated with arrays of micrococoons from silicon oxide nanowires by vapor deposition of a fluoropolymer. In the second case, a hydrophobic coating was deposited on heater surfaces with cavity microstructures formed by laser ablation and chemisorption of fluorinated methoxysilane. Water under saturation conditions at atmospheric pressure was used as the working liquid. The temperature of the heating surface was varied from 100 to 125 °C, and the maximum value of the heat flux was 160 W/cm^2. Boiling heat transfer on the test biphilic surfaces was significantly (up to 600%) higher than on non-biphilic surfaces. Surface texture, the shape of hydrophobic regions, and the method of their creation tested in this study did not show a significant effect on heat transfer. The boiling heat transfer rate was found to depend on the size of hydrophobic spots, the distance between them, and hence the number of spots. The highest heat transfer efficiency was detected for the surface with the largest number of hydrophobic spots. After long-term experiments (up to 3 years), the heat transfer coefficient on the obtained surfaces remained higher than on the smooth copper surface. Biphilic surfaces with arrays of cavities formed by laser ablation turned out to be the most stable during prolonged contact with boiling water.

Keywords: wettability; biphilic; surface modification; boiling; heat transfer enhancement

Citation: Chinnov, E.A.; Khmel, S.Y.; Vladimirov, V.Y.; Safonov, A.I.; Semionov, V.V.; Emelyanenko, K.A.; Emelyanenko, A.M.; Boinovich, L.B. Boiling Heat Transfer Enhancement on Biphilic Surfaces. *Energies* **2022**, *15*, 7296. https://doi.org/10.3390/en15197296

Academic Editor: Marco Marengo

Received: 3 September 2022
Accepted: 29 September 2022
Published: 4 October 2022

1. Introduction

Boiling heat transfer makes it possible to remove high heat fluxes at low overheating of the heated surface relative to the saturated vapor temperature because of the high latent heat of the phase transition. Boiling is widely used to cool electronic devices, nuclear reactors, electric power generators, etc. Biphilic surfaces can be suitable for cooling smooth surfaces of chips and optical devices of high-power lasers.

Boiling is a complex process that depends on many factors, in particular, the roughness (structure) and wettability of the surface on which this process occurs. Recently, there has been an increase in boiling studies associated with the development of new micro/nano surface modification techniques that provide heat transfer enhancement [1–7]. Micro/nano modification includes both micro/nano surface structuring and application of thin coatings (nanocoatings). Both approaches change the physicochemical properties of boiling surfaces, in particular, wettability.

It is known that nucleate boiling begins earlier on hydrophobic surfaces than on hydrophilic surfaces because hydrophobic surfaces have more centers of vapor bubble nucleation. This is due to the higher gas concentration in submicron cavities on hydrophobic surfaces; in addition, after bubble detachment, a larger amount of vapor remains on hydrophobic surfaces compared to hydrophilic surfaces [8]. As a result, at low heat fluxes, the boiling heat transfer coefficient on a hydrophobic surface is higher than on a hydrophilic surface. With an increase in the heat flux, the number of bubbles increases and they begin to coalesce, which prevents liquid access to the surface and deteriorates heat transfer, leading to the appearance of dry spots and, ultimately, to a boiling crisis. Boiling on hydrophilic surfaces starts later than on hydrophobic surfaces; however, at increased heat fluxes, the boiling heat transfer coefficient of hydrophilic surfaces is higher due to good surface wettability, leading to an enhanced liquid supply to the surface. For the same reason, the critical heat flux is higher. Thus, the use of hydrophobic surfaces is preferred at low heat fluxes, and the use of hydrophilic surfaces at high heat fluxes.

Several approaches have been proposed to simultaneously utilize the advantages of both types of surfaces in boiling processes. One approach is to use surfaces with mixed (heterogeneous, patterned) wettability or biphilic surfaces. Although this approach has been proposed previously [9,10], increased interest of researchers in the effect of biphilic surfaces on pool boiling heat transfer enhancement arose after the publication of a study [11] in 2010. Apparently, this is due to the above-mentioned development of new micro/nano surface modification techniques. These techniques emerged primarily in microelectronics and MEMS and were developed for silicon. Therefore, a significant part of the studies of biphilic surfaces was carried out on silicon substrates [8,11–14]. Hydrophilic surfaces were obtained, as a rule, by oxidation of silicon substrates [8,11–14]. Hydrophobic surfaces were obtained by applying a fluoropolymer coating using the spin-coating method [11–14]. Particle deposition can be analyzed using the method presented in [15]. Surface patterns were made by photolithography.

Pool boiling experiments have shown [11] an increase in both the heat transfer coefficient and the critical heat flux on surfaces with hydrophobic spots (regions) located on a hydrophilic background (surrounding) surface. Apparently, a large number of vaporization centers are formed on the hydrophobic surface of the spots, which provides rapid boiling at low overheating and a high heat transfer coefficient. In turn, the background hydrophilic surface between hydrophobic spots prevents bubble coalescence and increases the critical heat flux. In addition, the liquid and vapor flows are separated, which facilitates liquid access to the surface and contributes to an increase in the critical heat flux. Regular hexagonal spots were formed at the nodes of a uniform square lattice. The distance between the centers of the spots (pitch) was varied from 50 to 200 μm, and the size of hydrophobic spots was 40–60% of this distance.

Jo et al. [12] also investigated pool boiling heat transfer on biphilic surfaces consisting of hydrophobic spots patterned on a hydrophilic background surface. The effect of the hydrophobic spot size (50–500 μm), spot pitch (0.5 and 1 mm), and the number of hydrophobic spots (150 and 600) was studied. It has been shown that at low heat fluxes (<30 W/cm^2), the heat transfer coefficient is determined by the size of hydrophobic spots and the distance between them. At high heat fluxes, it is determined by the number of spots and their size. It is argued that the area ratio (the ratio of the total hydrophobic surface area to the total boiling area) has little effect on the heat transfer coefficient.

It is known that surface roughness has a significant effect on wettability; in particular, an increase in roughness leads to an increase in both the hydrophilicity of hydrophilic surfaces and the hydrophobicity of hydrophobic surfaces [10,16]. Betz et al. [13] used this circumstance to create biphilic surfaces with a maximum wettability gradient. Superbiphilic surfaces exhibited exceptional pool boiling performance with high critical heat fluxes (>100 W/cm^2) and very high heat transfer coefficients (10 W/cm^2 K). In particular, this coefficient was more than twice the heat transfer coefficient for biphilic surfaces with contact angles of 20° and 120°.

Jo et al. [14] systematically studied the dependences of the critical heat flux and boiling heat transfer on biphilic surfaces on the size and number of hydrophobic dots, pitch, and area ratio. It was found that the critical heat flux is inversely proportional to the area ratio and reaches maximum values for small values of this ratio. Boiling heat transfer depends on the size and number of hydrophobic dots and the pitch between them, but weakly depends on the area ratio. Therefore, for a significant enhancement of pool boiling heat transfer, a biphilic surface should contain a large number of hydrophobic dots 50–100 μm in size and at the same time should have a relatively small total hydrophobic area.

Motezakker et al. [8] studied boiling parameters on biphilic surfaces for different area ratios. Maximum values of the critical heat flux and the heat transfer coefficient were achieved at an area ratio of 38.46%, which corresponded to a hydrophilic spot diameter of 700 μm.

Other common materials used to fabricate substrates (heaters) when creating biphilic surfaces are metals and, above all, copper [17–21]. In this case, the hydrophilic surface was a metal surface or a thin film of another material deposited on metal. Hydrophobic surfaces were obtained by applying a fluoropolymer coating or silicone. Surface patterns were made by photolithography or other methods using a mask.

Takata et al. [17] created two types of biphilic surfaces: in one case, superhydrophobic spots of a nickel–fluoropolymer composite were deposited on a clean copper surface; in the other case, fluoropolymer spots were deposited on a superhydrophilic surface of titanium oxide deposited on a copper surface. In both cases, an enhancement of water pool boiling heat transfer was achieved. The determining parameters were the hydrophobic spot size and the area ratio. It was shown that a decrease in the spot size at a constant pitch leads to an increase in the heat transfer coefficient. For the same area ratio, the heat transfer coefficient was the same regardless of the type of coating, the shape of spots, etc. The best results were obtained for surfaces with minimum spot size and area ratio.

In a study [18], hydrophobic spots were formed from a nickel–fluoropolymer composite on a copper surface. The heat transfer coefficient was shown to be higher for smaller spot diameters and smaller pitches between spots.

The authors of [19,20] applied an original approach to the fabrication of biphilic surfaces aimed at minimizing their production cost. Hydrophobic spots and stripes were formed by spreading a high-temperature silicone sealant on a copper surface through a template. The spots were 2 and 4 mm in size with a pitch of 6 mm. Boiling heat transfer enhancement was obtained, with the enhancement being higher on spots of smaller diameter. However, the aging problem for such surfaces was not discussed in these papers.

Može et al. [21] created superbiphilic surfaces on an aluminum surface. Superhydrophobic spots were produced by CVD deposition of fluorinated silane on a nanostructured metal surface, and a superhydrophilic background (surrounding) surface was obtained by laser texturing. The heat transfer was found to significantly depend on the pitch and area ratio but weakly on the spot size. The maximum heat transfer coefficient was obtained for a superbiphilic surface with a spot size of 0.5 mm, a pitch of 1 mm, and an area ratio of 23%.

Promising methods for producing biphilic surfaces are laser techniques. These methods use beams with a minimum diameter of tens of microns. Therefore, a biphilic pattern for heat transfer enhancement can be easily achieved by scanning a substrate without using photolithography or other mask techniques. This approach was employed in [21,22], where certain surface areas were laser textured to obtain a biphilic pattern, and in [23], the wettability of the applied coating was changed by laser beam heating, resulting in a biphilic pattern.

Thus, experiments with silicon, aluminum, and copper substrates (heaters) have shown that the use of biphilic surfaces consisting of hydrophobic spots on a hydrophilic background surface leads to an enhancement of pool boiling heat transfer. Most studies have been performed on silicon substrates; studies on metal substrates, e.g., copper substrates, are markedly fewer in number. In particular, metal substrates with small hydrophobic spot sizes (less than 250 μm) have not been investigated, although it follows from the literature that the heat transfer enhancement is greater for smaller spot sizes. At the same time, the findings of the previous studies are contradictory: for example, in [12,14], it is stated that the heat transfer coefficient (HTC) depends weakly on the area ratio, whereas in [8,21], the optimal values of the area ratio are given, but they are different: 38.46% and 23%, respectively. We also note that the aging problem is not considered in papers on biphilic surfaces, although it is fundamental, as for any micro/nanomodified surface [24,25]. Further improvement in boiling heat transfer on biphilic surfaces can be achieved by preliminary micro/nanostructuring [8,13,26]; in particular, laser methods are promising for micro/nanostructuring [21]. New interesting approaches to enhancing boiling heat transfer have also been proposed—for example, the simultaneous use of a biphilic surface and nanofluids. [27].

In this paper, we present the results of a study of pool boiling heat transfer on biphilic copper surfaces with small hydrophobic spots. Spots of different shapes with an equivalent diameter of 50 to 100 μm were used. Boiling was investigated on three types of such surfaces with different micro/nanostructuring (micro/nanotexturing): a smooth copper surface, a copper surface coated with micro/nanostructures in the form of arrays of micrococoons from silicon oxide nanowires, and a copper surface with cavity microstructures formed by laser ablation. In addition, long-term experiments were carried out to study the problem of aging of biphilic surfaces.

2. Materials and Methods

2.1. Biphilic Surfaces and Their Fabrication Process

In this work, we used copper heaters with surface texture of three types: a smooth surface, a surface with micro/nanostructures in the form of arrays of micrococoons from silicon oxide nanowires, and a surface with arrays of microcavities formed by laser ablation. Heaters with biphilic surfaces were fabricated by two methods. In one method, a fluoropolymer coating was applied to the surface of heaters of the first and second types through a mask using hot wire chemical vapor deposition (HWCVD) with the formation of hydrophobic spots. In the other method, a hydrophobic coating was applied to the surface of heaters of the third type by chemisorption of fluorinated methoxysilane from the vapor phase at a temperature of 100–110 °C [28,29]. Detailed information on the characteristics of the surfaces used is given in Table 1.

Table 1. Characteristics of the surfaces used.

Surface Number	Type of Surface Texture	Hydrophobic Coating	Shape and Size of Hydrophobic Spots	Spot Pitch and Number of Spots	Area Ratio
1	Micrococoons from SiOx nanowires	No	No	No	0
2	Micrococoons from SiOx nanowires	Fluoropolymer HWCVD	0.1 mm diameter circle	0.5 mm, 69	2.8%
3	Micrococoons from nanowires SiOx	Fluoropolymer HWCVD	0.1 mm diameter circle	0.5 mm, 69	2.8%
4	Smooth copper surface	Fluoropolymer HWCVD	0.1 mm diameter circle	0.5 mm, 69	2.8%
5	Micrococoons from SiOx nanowires	Fluoropolymer HWCVD	0.05 mm diameter circle	0.2 mm, 506	5.1%
6	Smooth copper surface	Fluoropolymer HWCVD	0.05 mm diameter circle	0.2 mm, 506	5.1%
7	Copper surface with an array of cavities formed by laser ablation	Hydrophobization by chemisorption of fluorinated methoxysilane	Triangular cross-section cavity with an effective diameter of 0.07 mm	0.78 mm, 32	2.1%
8	Copper surface with an array of cavities formed by laser ablation	Hydrophobization by chemisorption of fluorinated methoxysilane	Rectangular cross-section cavity with an effective size of 0.1 mm	0.78 mm, 32	3.4%

A detailed description of the HWCVD for a fluoropolymer coating is given in [30,31]. We used two masks made of stainless steel with holes 50 μm and 100 μm in diameter located at the nodes of a square lattice with a pitch (distance between the centers of the holes) of 200 μm and 500 μm, respectively. These masks were tightly pressed against the heater surface during the deposition of the fluoropolymer coating. As a result, round hydrophobic spots 50 μm in diameter with a pitch of 200 μm and 100 μm in diameter with a pitch of 500 μm were formed on a smooth copper surface.

The second type of surface texture was obtained using arrays of micrococoons from silicon oxide nanowires. A micrococoon is a submicron-sized tin catalyst particle covered on all sides with silicon oxide nanowires. Tin catalyst particles were formed from a tin island film deposited by vacuum thermal evaporation. Before the deposition of tin, the copper heater surface was coated with a barrier layer of tungsten about 100 nm thick. Silicon oxide nanowires on tin catalyst particles were synthesized by gas-jet electron beam plasma chemical vapor deposition. The procedure for synthesizing arrays of micrococoons from SiO_x nanowires on copper heater surfaces is described in detail in [25]. SiO_x was chosen for the synthesis due to its high chemical and thermal stability. Such micro/nanostructures show the greatest resistance to destruction by boiling compared to other structures [25,31]. Figure 1a shows the surface morphology with arrays of micrococoons from silicon oxide nanowires before the deposition of the hydrophobic layer (fluoropolymer film). In Table 1, this surface is numbered 3.

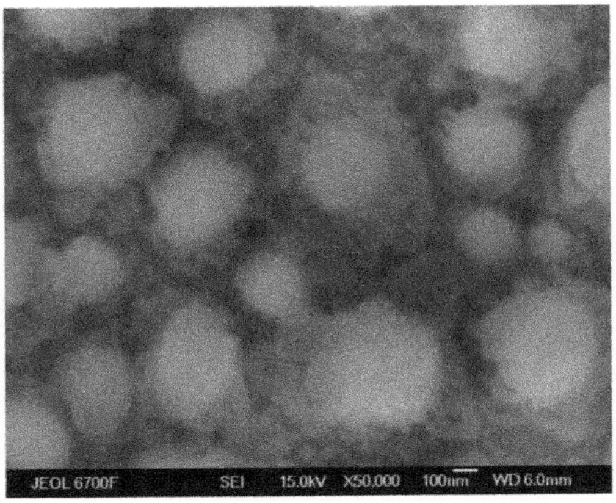

(a)

(b)

Figure 1. Surface morphology No. 3 before (**a**) and after the deposition of fluoropolymer (**b**).

Fluoropolymer film was deposited on these heaters by HWCVD, as in the previous case, using the same masks and under the same conditions. As a result, round hydrophobic spots 50 μm in diameter with a pitch of 200 μm and 100 μm in diameter with a pitch of 500 μm were formed on the copper heater surface with arrays of micrococoons. Figure 1b shows the surface morphology with arrays of micrococoons after the deposition of the fluoropolymer film. Features of the formation of hydrophobic spots on arrays of micrococoons from silicon oxide nanowires are given in [32]. Figure 2 shows SEM images at different magnifications of the surface after the deposition of fluoropolymer spots. It can be seen that the deposition through the mask results in round fluoropolymer spots.

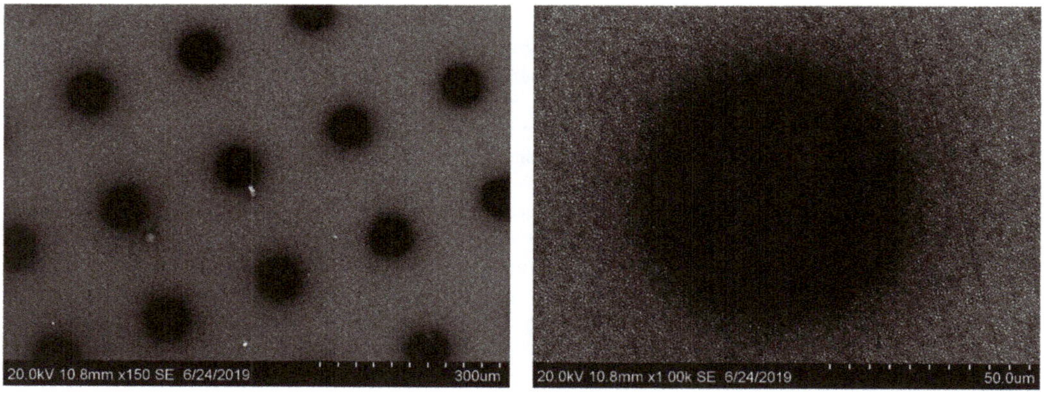

(**a**) (**b**)

Figure 2. SEM images of surface No. 5 after the deposition of the fluoropolymer coating through a 50–200 µm mask (hole diameter 50 µm, pitch 200 µm): (**a**) magnification 150, scale bar 300 µm, (**b**) magnification 1000, scale bar 50 µm.

The elemental composition of these coatings was analyzed by EDS. Figure 3 shows surface mapping images for different elements. Fluorine and carbon (fluoropolymer components) are clearly detected in the region of spots (Figure 3a,b), and silicon dominates in the region between the spots (Figure 3c). This confirms that the composition of the spots corresponds to the fluoropolymer and that there is no fluoropolymer in the space between the spots.

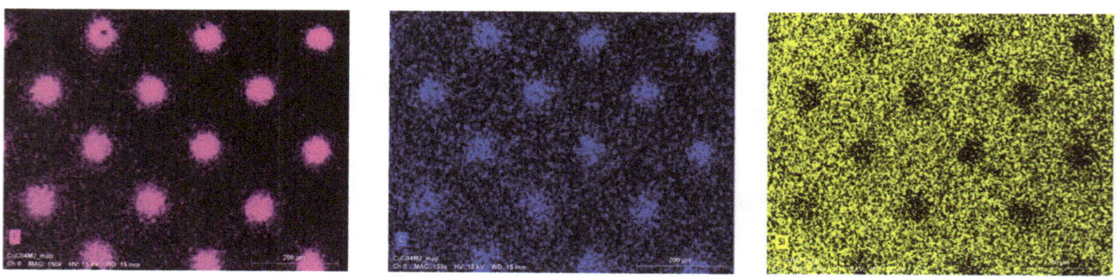

(**a**) (**b**) (**c**)

Figure 3. Elemental composition map of the biphilic heater surface obtained using a 50–200 µm mask: (**a**) fluorine, (**b**) carbon, and (**c**) silicon. Scale bar: 200 µm.

The third type of surface texture was obtained by laser ablation with the formation of arrays of cavities with an equivalent diameter of about 70–100 µm on the boiling surface. For texturing, we used an Argent-M laser marking system (Laser Technology Center, Russia) based on an infrared ytterbium fiber laser with a wavelength of 1.064 µm, a focused laser spot diameter (at the 1/e2 level) of 40 µm, and a peak power of up to 0.95 mJ in TEM00 mode. Local surface regions were subjected to a single or multiple laser treatment with a pulse duration of 200 ns and a pulse frequency of 20 kHz. Surfaces with arrays of cavities were fabricated. In the single pass mode, 1568 pulses were used to fabricate one cavity and the resulting cavities had a triangular cross section. In the multiple pass mode, 7840 pulses were used and rectangular cross-section cavities were obtained. Figure 4 shows images of individual triangular and rectangular cavities.

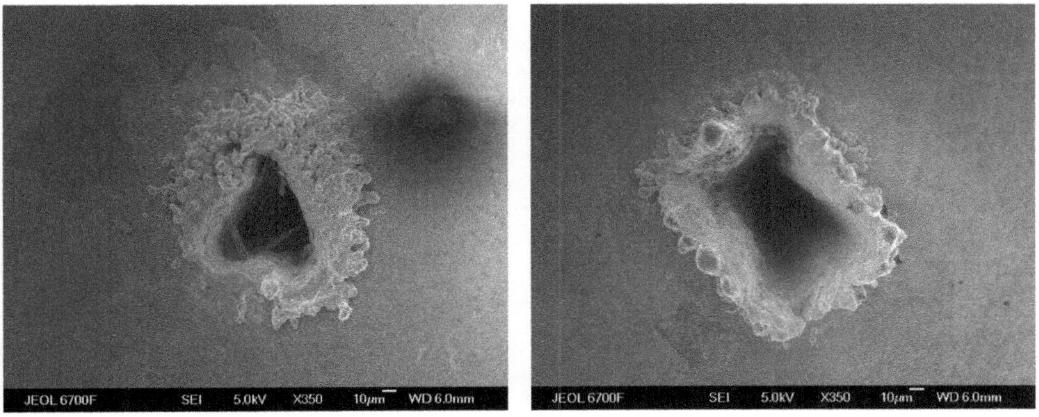

(a) (b)

Figure 4. SEM image of an individual cavity on surface No. 7 (**a**) and surface No. 8 (**b**). Scale bar: 10 μm.

The cross-section of the triangular cavity has an equivalent diameter of about 70 μm, and the cross-section of the rectangular cavity has an approximate size of 100 μm. For both surfaces, the pitch was 780 μm. To create biphilic surfaces, copper surfaces were hydrophobized by UV/ozone activation, followed by chemisorption of fluorinated methoxysilane from the vapor phase at a temperature of 100–110 °C. This resulted in the formation of a layer of two-dimensional chemically cross-linked fluoroxysilane which is bonded through hydroxyl groups to the surface at ablation sites [28,29]. It was expected that the hydrophobized layer on the smooth copper surface would be quickly destroyed, whereas in laser-treated areas it will remain for a long time. The main factor in this process should be different types of adsorption of fluoroxysilane molecules: stronger chemical adsorption in laser-treated areas and less strong physical adsorption in untreated areas. Thus, a biphilic surface is formed.

Figure 5 shows SEM images at different magnifications of the side wall of the rectangular cavity. It can be seen that the laser-treated surface becomes rough and porous.

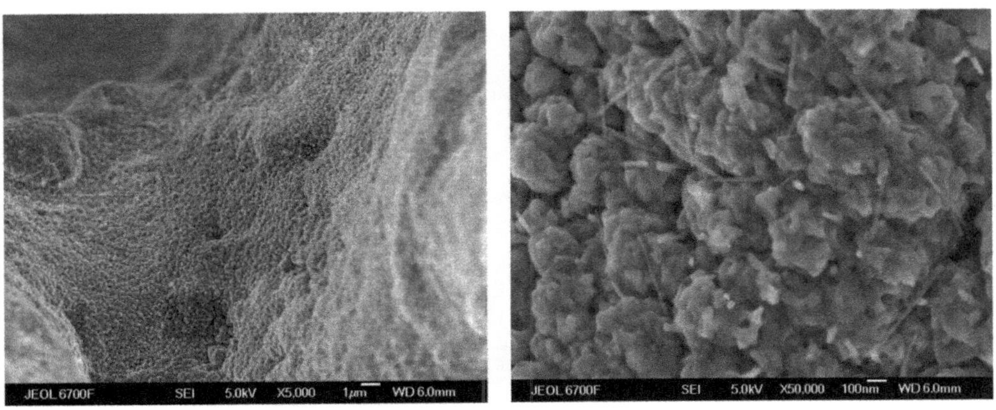

(a) (b)

Figure 5. SEM image of the wall of an individual cavity on surface No. 8 at different magnifications. Scale bar: (**a**) 1 μm, 5000 magnification; (**b**) 100 nm, 50,000 magnification.

The contact angle for the smooth copper surface was 54 ± 2°, and that for the copper surface with the deposited fluoropolymer was 140 ± 2°. The contact angle for the array of micrococoons from silicon oxide nanowires was 36 ± 2°, and after fluoropolymer deposition, it was 164 ± 2°. For hydrophobized with fluorinated methoxysilane samples, the contact angle was about 170 ± 2° in ablated regions [32]. The roughness R_a of the smooth copper surface was no more than 0.17 μm.

The characterization of the samples was carried out using a JEOL JSM-6700F and a Hitachi S-3400N scanning electron microscopes equipped with energy dispersive X-ray spectrometers (EDS) for element analysis. The water contact angle was measured using a KRUSS DSA-100 device.

2.2. Experimental Setup

The experimental setup is shown schematically in Figure 6 with an indication of the main components. The main component of the setup is a thermostated chamber (see Figures 7 and 8) with five optical windows for monitoring the processes occurring during the experiments. The walls of the chamber are parallel steel plates, between which liquid circulates. The chamber consists of three independent circuits: a box with side walls, an upper wall (lid), and a lower wall (bottom). Each circuit includes a system of connecting pipes and valves and is connected to a thermostat to maintain a uniform and constant temperature distribution on the walls of the chamber. The spatial temperature distribution is monitored by two thermocouples brought into the chamber through a common sealed port and mounted on each wall. The chamber is filled to the required level with the working liquid through the liquid supply channel. At the bottom of the chamber there is a cylindrical working section with a copper core fixed with a special hold-down device. Thermocouples used to measure and calculate the heat flux from the heater are distributed over the working section.

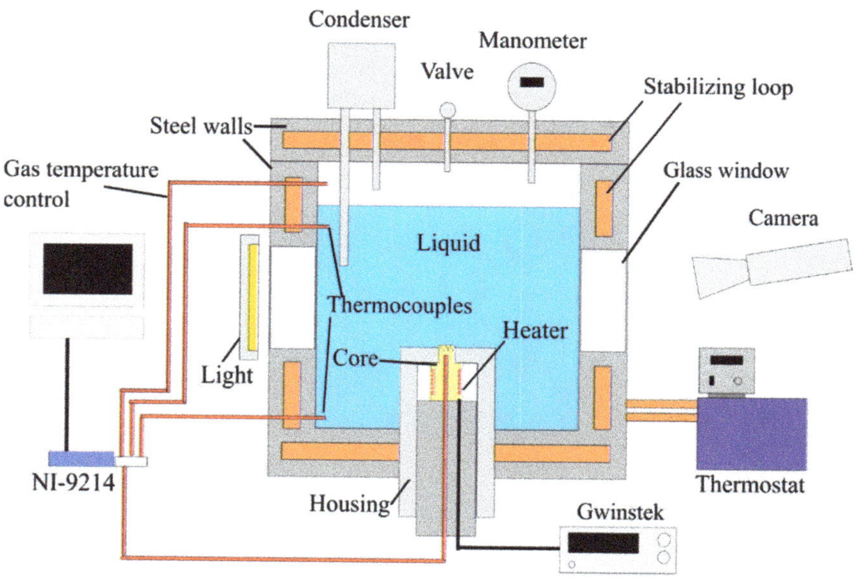

Figure 6. Sketch of the experimental setup.

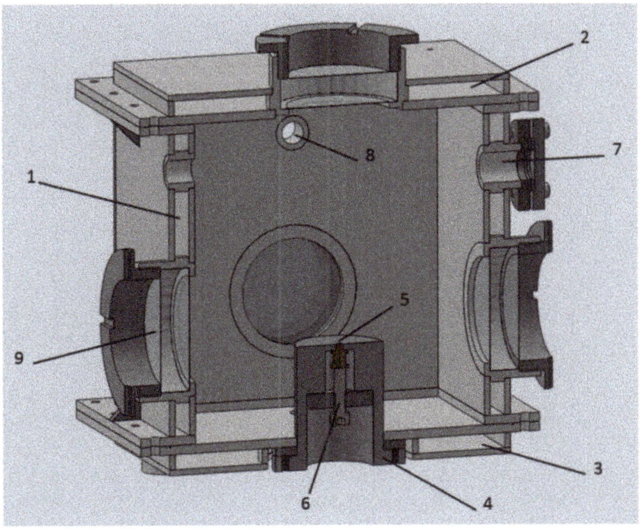

Figure 7. View of the thermostated chamber: 1—lateral thermal stabilization circuit, 2—upper thermal stabilization circuit, 3—lower thermal stabilization circuit, 4—working section, 5—heating element, 6—hold-down device, 7—thermocouple entry port, 8—sensor entry port, and 9—optical window.

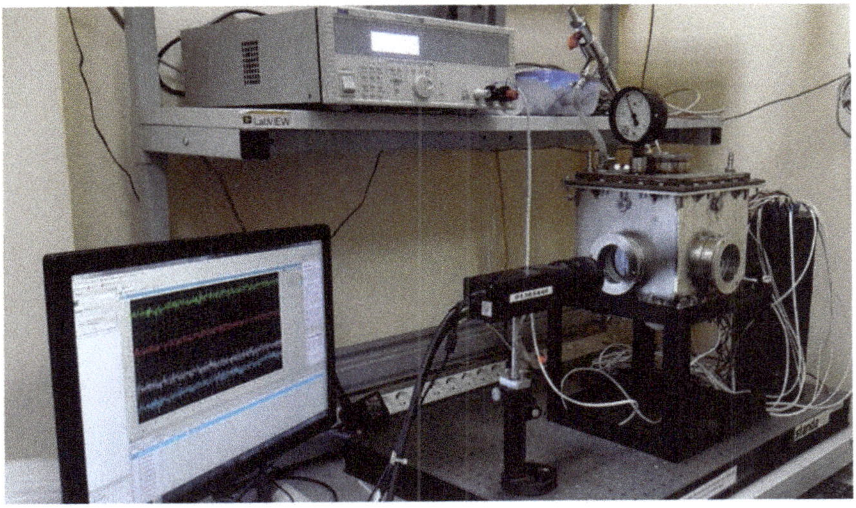

Figure 8. View of the experimental setup.

On the upper wall of the chamber, there is a finned tube steam condenser connected to a cooling circuit with a thermostat. A steam pressure sensor and a pressure relief valve are also located in the upper part of the chamber. A high-speed video camera for high-precision recording of the processes under study is placed opposite the side optical window. An illumination system is mounted on the other side of the thermostated chamber to provide sufficient lighting conditions of the working section during high-speed shooting. An additional general view camera for visualization and monitoring of the experiment is placed above the upper optical window.

The instrument and control system consisted of an NI-9214 data acquisition system and software. The temperature measurement sensitivity of this unit is 0.1 °C. The temperature of the heater, liquid, and environment was measured using thermocouples. Thermocouples were connected with a cold junction (no automatic cold junction compensation) to improve accuracy. The temperature of the heater, liquid, and environment is measured using thermocouples. All thermocouples of the setup were individually calibrated from 0 °C to 150 °C with an interval of 10 °C. A KS 100-1 dry-well calibrator was used below 100 °C, and a Termex VT-8 glycerin thermostat in the range from 100 °C to 150 °C. In both cases, temperature was monitored using a V7-99 meter with ETS-100 resistance thermometers. The calibration error was 0.1 °C. Distilled, deionized, and degassed water (Milli-Q) was used as the working liquid.

The synthesis and deposition of nanocoatings by different methods requires the use of sufficiently small heaters that can be placed in synthesis and deposition chambers, quickly mounted into the setup without damage to the nanostructures on the end heating surface, removed, and regularly monitored with an electron microscope. To perform experiments under the same carefully controlled conditions, it is necessary to fabricate a large series of heaters with identical characteristics.

Comparison of our method with other methods, e.g., gluing silicon wafers with nanostructures to a heater, showed the advantages of the approach used. The heating element is a cylindrical copper core with a head of 5 mm diameter (Figure 9). The boiling surface is the flat top end of the cylinder. The tip of the core is tightly inserted into the hole of the fluoropolymer base and is aligned in one plane with the top of the base. The heat source is a Nichrome wire tightly wound around the core stem with a resistance of 4 Ohm. The heating element is carefully insulated to minimize heat loss. Glass fabric wrapped in several layers around the core of the heating element is used as a heat insulator.

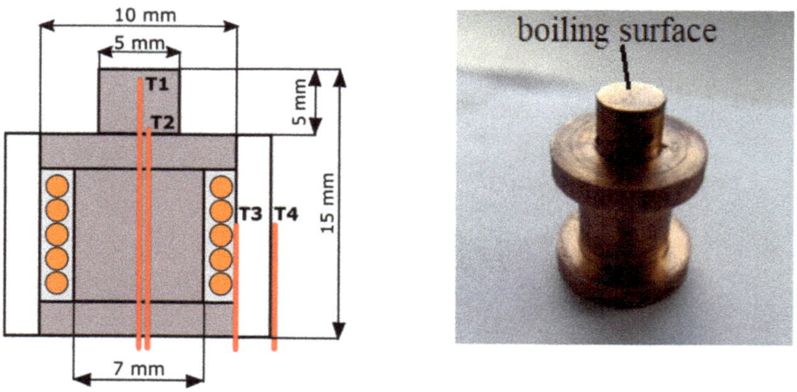

(**a**) (**b**)

Figure 9. Sketch (**a**) and photo (**b**) of the typical heating element (photo was created before modifications).

Two thermocouples T1 and T2 were mounted in grooves at different depths in the copper core (Figure 9) to measure the surface temperature and determine the heat flux according to the Fourier law. To monitor the heat flux distribution, additional thermocouples T3 and T4 were mounted in the region of the core and used for winding the Nichrome heater.

The heat flux supplied to the test section was calculated by Fourier's law [33] as:

$$q = -k_{Cu}\frac{T2 - T1}{X1} \tag{1}$$

where q is the heat flux, $X1$ is the distance between thermocouples $T2$ and $T1$, and k_{Cu} is the thermal conductivity of copper.

The surface temperature was determined from the equation [33]:

$$T_w = T1 - q\left(\frac{X2}{k_{Cu}}\right) \quad (2)$$

where T_w is the temperature of the test surface and $X2$ is the distance between the test surface and thermocouple $T1$.

The main source of the measurement error was inaccuracy in determining the real heat flux. This inaccuracy is due to the calibration error of thermocouples $T1$ and $T2$, the error in determining the depth of grooves for thermocouple placement, and the inaccuracy of their positioning in the grooves. The calibration error was 0.1 °C, as already mentioned above. The total positioning error of the thermocouples was about 0.01 mm. The measurement error was estimated using the well-known formulas mentioned in [33]. The final error of temperature determination was no more than 0.2 °C. The heat flux error did not exceed 3 W/cm^2.

3. Results

3.1. Validation of the Experimental Setup and Measurement Technique

Before the main experiments, a series of test measurements was performed on smooth copper surfaces. The results of the series for the three experiments of pool boiling on smooth copper are shown in Figure 10.

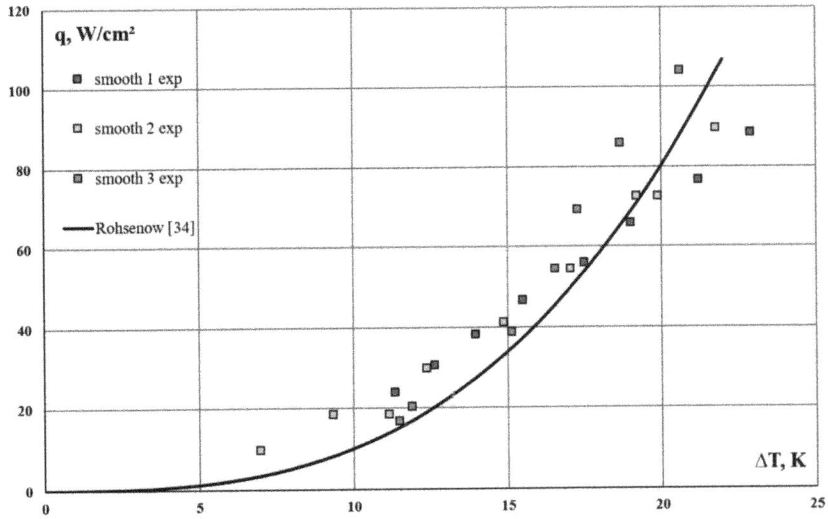

Figure 10. Dependence of the heat flux on the temperature difference between the heating surface and the liquid under saturation conditions for the smooth surface.

The well-known Rohsenow correlation [34] was used for comparison. For the correlation, the coefficients for surfaces with the closest characteristics were used. The data are in good agreement with the theoretical curve. The average deviation does not exceed 15%. The second experiment has the least standard deviation (8 W/cm^2) from the theoretical curve. These results exceed the previously calculated experimental errors but this can be explained by the incompletely matched surface parameters (roughness, degree of oxidation). Thus, the experimental setup and the measurement technique can be considered valid and to provide reliable results. Further, all experimental results are compared with the Rohsenow correlation.

3.2. Experimental Results

The effect of the size of round hydrophobic spots and the pitch between them on the boiling heat transfer rate was studied. In the first series of experiments, surfaces with hydrophobic fluoropolymer regions 100 μm in diameter and a pitch of 500 μm were used. A boiling curve was first obtained for a smooth copper surface with fluoropolymer spots (surface No. 4). The result is shown in Figure 11. It can be seen from the figure that the heat transfer on the biphilic surface is enhanced compared to the homogeneous smooth copper surface (Rohsenow correlation). A similar result was obtained for surfaces covered with arrays of micrococoons: boiling heat transfer on surfaces No. 2 and No. 3 with fluoropolymer spots is more intense than on surface No. 1 (a homogeneous array of micrococoons without hydrophobic spots). Note that the results for surfaces No. 2 and No. 3 are approximately the same, which indicates that the technology is reproducible. Thus, biphilicity enhances heat transfer on both smooth surfaces and surfaces covered with micrococoons from silicon oxide nanowires. It is additionally notable that the results on smooth surface No. 4 are slightly superior to the results on surfaces No. 2 and No. 3, despite the initial assumption that pre-coating with micrococoons would enhance boiling heat transfer on the biphilic surface.

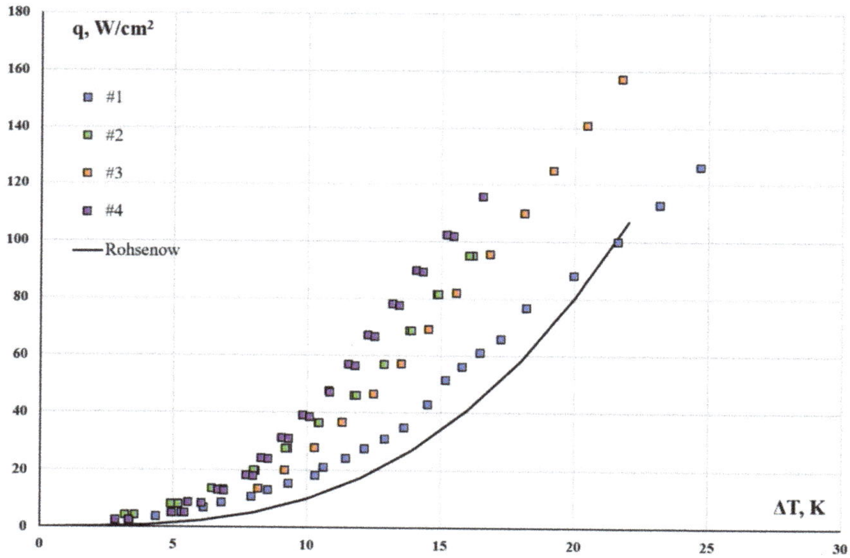

Figure 11. Boiling curves for surface No. 1 and surfaces Nos. 2–4 from Table 1 with deposited fluoropolymer spots 100 μm in diameter with a pitch of 500 μm.

Surfaces with hydrophobic fluoropolymer regions 50 μm in diameter with a pitch of 200 μm exhibit a similar behavior in experiments (Figure 12). The heat transfer performance of smooth copper surface No. 6 with hydrophobic spots is higher than that of surface No. 5 covered with arrays of micrococoons with hydrophobic spots. However, in this case, the effect is much less pronounced and is observed at heat fluxes of about 60 to 130 W/cm^2.

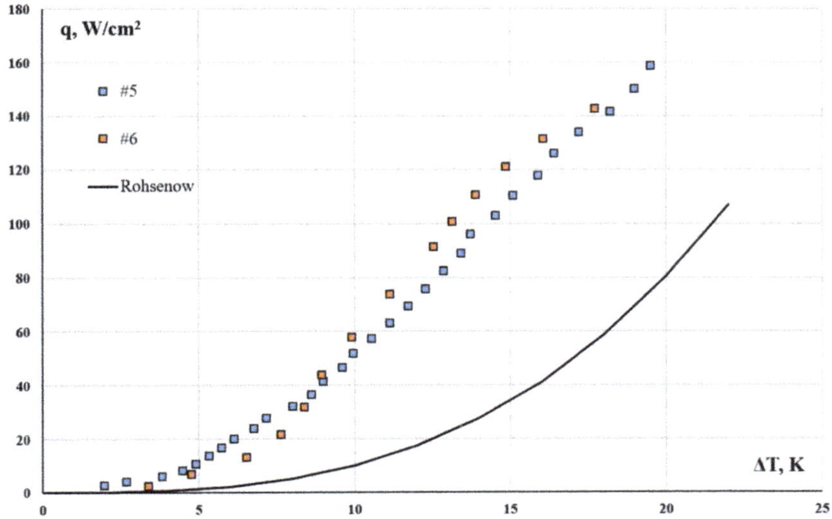

Figure 12. Boiling curves for surface Nos. 5 and 6 from Table 1 with deposited fluoropolymer spots 50 μm in diameter with a pitch of 200 μm.

A comparison of the efficiency of using surfaces with fluoropolymer spots 100 μm in diameter with a pitch of 500 μm and surfaces with spots 50 μm in diameter with a pitch of 200 μm shows a fairly significant increase of up to 20% in heat transfer rate for surfaces with smaller spots and a smaller pitch. Moreover, this effect is observed both for surfaces with arrays of micrococoons (Nos. 2, 3, and 5) (Figure 13) and for smooth copper surfaces (Nos. 4 and 6) (Figure 14).

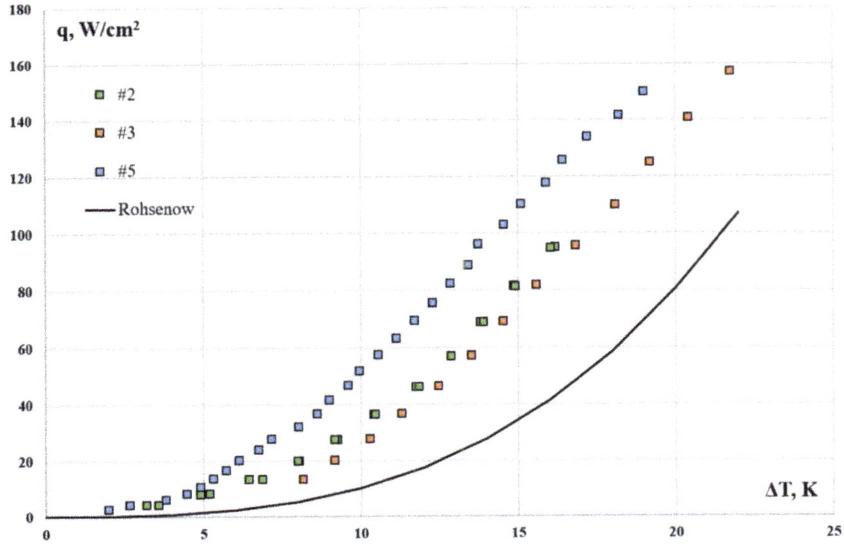

Figure 13. Comparison of boiling curves for surfaces Nos. 2 and 3 (spot diameter 100 μm, pitch 500 μm) and surface No. 5 (spot diameter 50 μm, pitch 200 μm) with arrays of micrococoons.

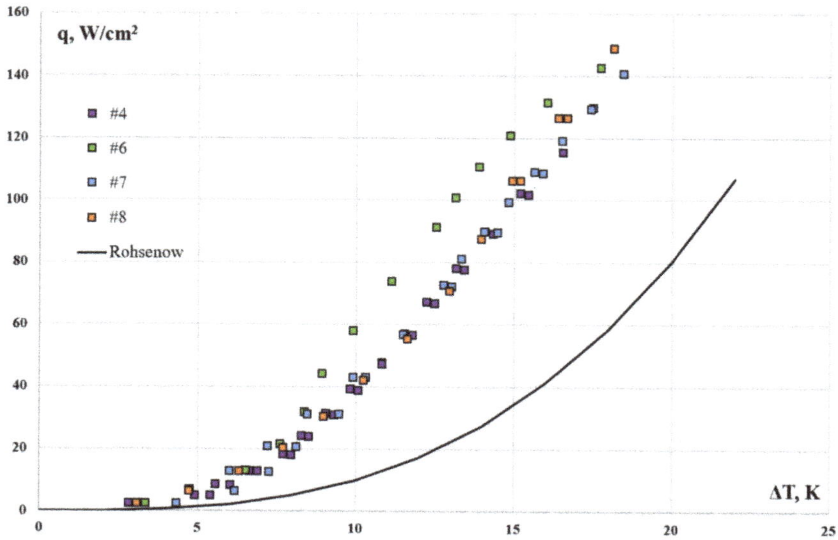

Figure 14. Boiling curves for surfaces Nos. 4, 6, 7, and 8.

Figure 14 shows data for surfaces No. 4 (spot diameter 100 μm, pitch 500 μm), No. 6 (spot diameter 50 μm, pitch 200 μm), and for surfaces Nos. 7 and 8 (triangular and rectangular cavities, respectively) obtained by laser methods.

It can be seen that the heat transfer rate on surface No. 6 is 20% higher than on the other surfaces, including surface No. 4. Thus, in view of the data in Figure 13, it can be concluded that for all biphilic samples with fluoropolymer spots of 50 μm in diameter and a pitch of 200 μm, the heat transfer efficiency is higher than for samples with spots of 100 μm in diameter and a pitch of 500 μm.

4. Discussion

4.1. Effect of Surface Properties on Boiling

All biphilic surfaces show an intensity of heat transfer that is more than twice as high (up to 600%) as a smooth copper surface. Heat transfer intensification on surfaces obtained using a 50/200 mask (Nos. 5 and 6) is 20% higher than on the other surfaces.

Biphilic surfaces Nos. 7 and 8 are most comparable in the spot area (in the plane of the surface) and spot pitch to surface No. 4. The heat transfer rates on these three surfaces are almost identical, although their shape, texture, and depth of cavities are significantly different. Apparently, the initial assumption that the hydrophobic layer is rapidly destroyed on the main surface and has good stability in cavities obtained by ablation turns out to be valid.

It was already mentioned above that the boiling heat transfer rate on biphilic surfaces is significantly affected by the following factors: the size of hydrophobic spots, the distance between them (pitch), and the area ratio (the ratio of the total hydrophobic surface area to the total boiling area). A significant difference of surface No. 6 is that it has a much larger (by an order of magnitude) number of hydrophobic spots: 506 versus 69 for surface No. 4 and 32 for surfaces No. 7 and No. 8. The larger area ratio for surface No. 6 (5.1%) compared to the area ratio for surfaces No. 4 (2.8%), No. 7 (2.1%), and No. 8 (3.4%) may also affect the heat transfer enhancement.

It can be assumed that the use of a small number of large spots is indeed less effective since closely spaced vaporization centers in the same spot interfere with each other, and it is worth isolating them from each other by reducing the spot size in parallel with decreasing the spot pitch. However, in general, the question of the optimal characteristics of biphilic surfaces remains open and requires further detailed study.

4.2. Comparison with Literature Data

A review of the literature showed that the available data on pool boiling of water on flat copper heaters at atmospheric pressure are few in number. Systemic data on the optimal geometry of hydrophobic regions and the optimal area ratio are absent or inconsistent. A comparison with the results of the studies cited above is shown in Figure 15. It can only be noted that our data on the degree of hydrophobicity are closest to those reported in the paper by Yamada et al. [18], where hydrophobic regions occupy approximately 2.2%. It can be seen that the data obtained for some of the most characteristic surfaces (Nos. 4, 6, and 8) are superior to the results of Yamada.

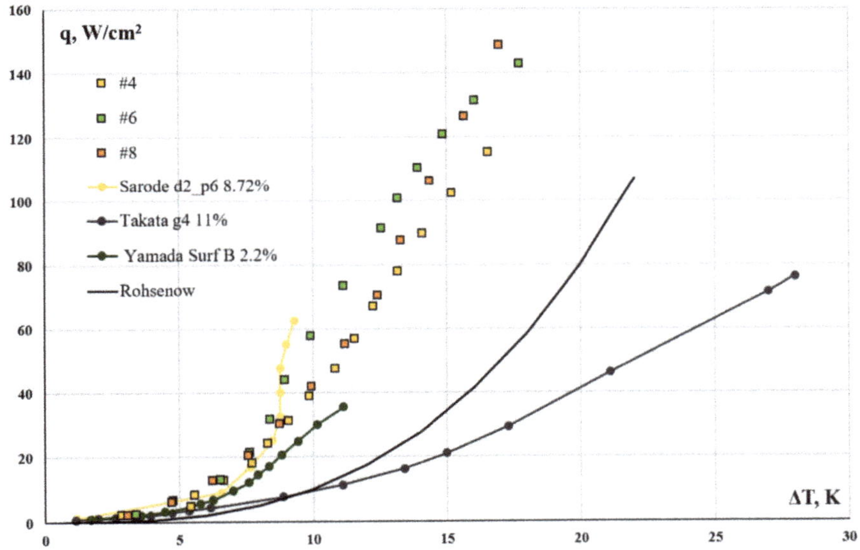

Figure 15. Boiling curves for surfaces Nos. 4, 6, and 8 and some literature data.

4.3. Stability of the Biphilic Surfaces

An important problem in the operation of various micro/nanomodified surfaces is their aging. During operation, many surfaces undergo morphological changes that can be associated with mechanical destruction by boiling due to the effect of growing bubbles on micro/nanostructures and/or due to different thermal expansion coefficients of surface materials during layer-by-layer deposition of films of several materials. In addition, intense hydrothermal loads can cause chemical changes on surfaces.

In the literature, insufficient attention is paid to these issues and there is no mention of the reproducibility of experiments.

In this work, we performed a series of long-term (two-year) experiments for surfaces Nos. 7 and 8 (with cavity microstructures formed by laser ablation), in which samples of these surfaces demonstrated high stability. As an example, Figure 16 shows boiling curves for surface No. 7 measured on different dates. It can be seen from these data that the pool boiling heat transfer rate of the surface remains high throughout the period of operation and the boiling curves do not show significant changes.

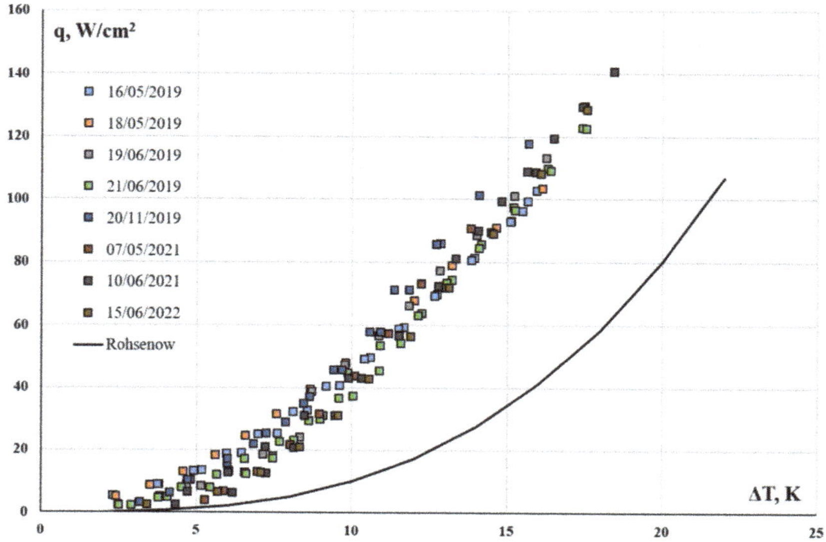

Figure 16. Boiling curves for surface No. 7 showing its stability. DD/MM/YYYY format was used.

Surfaces obtained by local deposition of the fluoropolymer through a mask onto smooth copper did not exhibit high stability compared to surfaces No. 7 and No. 8. As an example, Figure 17 shows some data for surface No. 4. After the best result obtained on 7 June 2019, the heat transfer rate decreased in the subsequent experiments.

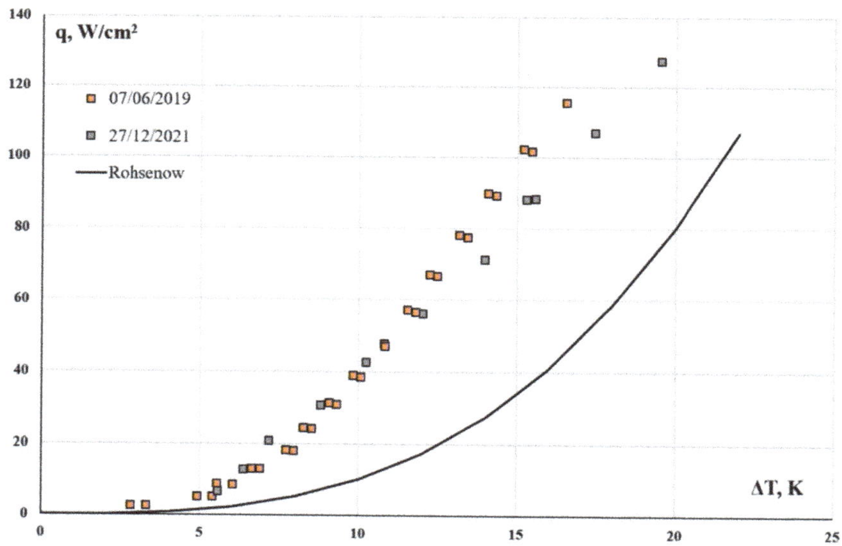

Figure 17. Boiling curves for surface No. 4.

Surfaces obtained by local deposition of the fluoropolymer on an array of micrococoons also did not exhibit high stability, despite the initial assumptions that the adhesion of the fluoropolymer deposited on microstructures would be higher than in the case of deposition on smooth copper. Some data for surface No. 3 are given in Figure 18. Just as for surface No. 4, the heat transfer rate decreased after the first experiments.

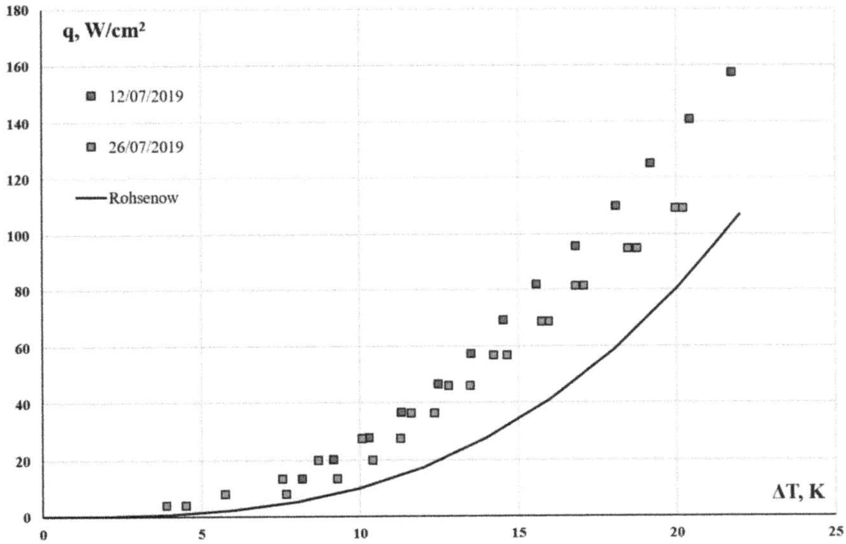

Figure 18. Boiling curves for surface No. 3.

The standard deviations between the experimental curves in both cases exceed 11 W/cm^2, which is much larger than the experimental error.

5. Conclusions

1. Boiling experiments were carried out using two types of biphilic surfaces obtained by different techniques. The degrees of heat transfer enhancement on nano-modified surfaces were determined. The effect of the size (0.05, 0.07, 0.1 mm), shape (circle, triangular, rectangular), and number of hydrophobic spots (32 to 506) on the boiling heat transfer rate on eight copper surfaces was studied.
2. Both techniques demonstrated their effectiveness and provided a significant enhancement of boiling heat transfer. The heat transfer coefficient on biphilic surfaces was significantly (up to 600 %) higher than on smooth surfaces or surfaces coated with cocoons.
3. The boiling heat transfer rate depended on the size of hydrophobic spots, the distance between them, and hence on their number. The results complement previous data [7,18] and show that heat transfer enhancement can be obtained on surfaces with a small diameter of hydrophobic regions (0.05 mm) and a relatively small area ratio (5%). At the same time, no significant effect of surface texture and shape of hydrophobic regions tested in this study on heat transfer was found.
4. After long-term experiments (over 1 to 3 years), the heat transfer coefficient on the obtained surfaces remained higher than on a smooth copper surface. During 3 years of operation, the best stability of enhanced heat transfer was observed for biphilic surfaces with cavities formed by laser ablation.

Author Contributions: Conceptualization, E.A.C. and S.Y.K.; funding acquisition, E.A.C.; investigation, V.Y.V. and V.V.S.; methodology, S.Y.K., A.I.S., K.A.E., A.M.E. and L.B.B.; project administration, E.A.C.; resources, S.Y.K., A.I.S., K.A.E., A.M.E. and L.B.B.; supervision, E.A.C.; validation, V.Y.V. and V.V.S.; visualization, S.Y.K. and V.Y.V.; writing—original draft, E.A.C., S.Y.K. and V.Y.V.; writing—review and editing, E.A.C., S.Y.K., V.Y.V., K.A.E., A.M.E. and L.B.B. All authors have read and agreed to the published version of the manuscript.

Funding: This work was funded by the Russian Science Foundation (grant number 22-19-20090) and the Government of the Novosibirsk Region (agreement number p-13).

Institutional Review Board Statement: Not applicable.

Informed Consent Statement: Not applicable.

Data Availability Statement: Not applicable.

Conflicts of Interest: The authors declare no conflict of interest. The funders had no role in the design of the study; in the collection, analyses, or interpretation of data; in the writing of the manuscript; or in the decision to publish the results.

Nomenclature

q—heat flux, W/cm^2; k_{Cu}—thermal conductivity of copper, W/(m*K); T—temperature, K, °C; X—distance, mm; Contact angle, °.

References

1. Liang, G.; Mudawar, I. Review of pool boiling enhancement by surface modification. *Int. J. Heat Mass Transf.* **2019**, *128*, 892–933. [CrossRef]
2. Li, W.; Dai, R.; Zeng, M.; Wang, Q. Review of two types of surface modification on pool boiling enhancement: Passive and active. *Renew. Sustain. Energy Rev.* **2020**, *130*, 109926. [CrossRef]
3. Chen, J.; Ahmad, S.; Cai, J.; Liu, H.; Lau, K.T.; Zhao, J. Latest progress on nanotechnology aided boiling heat transfer enhancement: A review. *Energy* **2021**, *215*, 119114. [CrossRef]
4. Kamel, M.S.; Lezsovits, F.; Hussein, A.K. Experimental studies of flow boiling heat transfer by using nanofluids A critical recent review. *J. Therm. Anal. Calorim.* **2019**, *138*, 4019–4043. [CrossRef]
5. Surtaev, A.S.; Serdyukov, V.S.; Pavlenko, A.N. Nanotechnologies for thermophysics: Heat transfer and crisis phenomena at boiling. *Nanotech. Russia* **2016**, *11*, 696–715. [CrossRef]
6. Dedov, A.V. A Review of Modern Methods for Enhancing Nucleate Boiling Heat Transfer. *Therm. Eng.* **2019**, *66*, 881–915. [CrossRef]
7. Serdyukov, V.; Patrin, G.; Malakhov, I.; Surtaev, A. Biphilic surface to improve and stabilize pool boiling in vacuum. *Appl. Therm. Eng.* **2022**, *209*, 118298. [CrossRef]
8. Motezakker, A.R.; Sadaghiani, A.K.; Çelik, S.; Larsen, T.; Villanueva, L.G.; Koşar, A. Optimum ratio of hydrophobic to hydrophilic areas of biphilic surfaces in thermal fluid systems involving boiling. *Int. J. Heat Mass Transf.* **2019**, *135*, 164–174. [CrossRef]
9. Hummel, R.L. Means for Increasing the Heat Transfer Coefficient between a Wall and Boiling Liquid. U.S. Patent No. 3207209, 21 September 1965. Available online: https://patents.google.com/patent/US3207209A/en (accessed on 5 July 2022).
10. Attinger, D.; Frankiewicz, C.; Betz, A.R.; Schutzius, T.M.; Ganguly, R.; Das, A.; Kim, C.-J.; Megaridis, C.M. Surface engineering for phase change heat transfer: A review. *MRS Energy Sustain.* **2014**, *1*, 4. [CrossRef]
11. Betz, A.R.; Xu, J.; Qiu, H.; Attinger, D. Do surfaces with mixed hydrophilic and hydrophobic areas enhance pool boiling? *Appl. Phys. Lett.* **2010**, *97*, 141909. [CrossRef]
12. Jo, H.; Ahn, H.S.; Kang, S.; Kim, M.H. A study of nucleate boiling heat transfer on hydrophilic, hydrophobic and heterogeneous wetting surfaces. *Int. J. Heat Mass Transf.* **2011**, *54*, 5643–5652. [CrossRef]
13. Betz, A.R.; Jenkins, C.-J.; Kim, C.-J.; Attinger, D. Boiling heat transfer on superhydrophilic, superhydrophobic, and superbiphilic surfaces. *Int. J. Heat Mass Transf.* **2013**, *57*, 733–741. [CrossRef]
14. Jo, H.J.; Kim, S.H.; Park, H.S.; Kim, M.H. Critical heat flux and nucleate boiling on several heterogeneous wetting surfaces: Controlled hydrophobic patterns on a hydrophilic substrate. *Int. J. Multiph. Flow* **2014**, *62*, 101–109. [CrossRef]
15. Colombo, L.; Alberti, L.; Mazzon, P.; Antelmi, M. Null-space Monte Carlo particle backtracking to identify groundwater tetrachloroethylene sources. *Front. Environ. Sci.* **2020**, *8*, 142. [CrossRef]
16. Wenzel, R.N. Resistance of solid surfaces to wetting by water. *Ind. Eng. Chem.* **1936**, *28*, 988–994. [CrossRef]
17. Takata, Y.; Hidaka, S.; Kohno, M. Effect of surface wettability on pool boiling: Enhancement by hydrophobic coating. *Int. J. Air-Cond. Refrig.* **2012**, *20*, 1150003. [CrossRef]
18. Yamada, M.; Shen, B.; Imamura, T.; Hidaka, S.; Kohno, M.; Takahashi, K.; Takata, Y. Enhancement of boiling heat transfer under sub-atmospheric pressures using biphilic surfaces. *Int. J. Heat Mass Transf.* **2017**, *115*, 753–762. [CrossRef]
19. Sarode, A.; Raj, R.; Bhargav, A. Scalable macroscale wettability patterns for pool boiling heat transfer enhancement. *Heat Mass Transf.* **2020**, *56*, 989–1000. [CrossRef]
20. Liang, G.; Chen, Y.; Wang, J.; Wang, Z.; Shen, S. Experiments and modeling of boiling heat transfer on hybrid-wettability surfaces. *Int. J. Multiph. Flow* **2021**, *144*, 103810. [CrossRef]
21. Može, M.; Zupančič, M.; Golobič, I. Pattern geometry optimization on superbiphilic aluminum surfaces for enhanced pool boiling heat transfer. *Int. J. Heat Mass Transf.* **2020**, *161*, 120265. [CrossRef]
22. Xia, Y.; Gao, X.; Li, R. Surface effects on sub-cooled pool boiling for smooth and laser-ablated silicon surfaces. *Int. J. Heat Mass Transf.* **2022**, *194*, 123113. [CrossRef]

23. Zupančič, M.; Steinbücher, M.; Gregorčič, P.; Golobič, I. Enhanced pool-boiling heat transfer on laser-made hydrophobic/superhydrophilic polydime thylsiloxane-silica patterned surfaces. *Appl. Therm. Eng.* **2015**, *91*, 288–297. [CrossRef]
24. Može, M. Effect of boiling-induced aging on pool boiling heat transfer performance of untreated and laser-textured copper surfaces. *Appl. Therm. Eng.* **2020**, *181*, 116025. [CrossRef]
25. Khmel, S.; Baranov, E.; Vladimirov, V.; Safonov, A.; Chinnov, E. Experimental study of pool boiling on heaters with nanomodified surfaces under saturation. *Heat Transf. Eng.* 2021. Available online: https://www.tandfonline.com/doi/full/10.1080/01457632.2021.2009211(accessed on 5 July 2022).
26. Cho, H.R.; Park, S.C.; Kim, D.; Joo, H.-m.; Yu, D.I. Experimental study on pool boiling on hydrophilic micro/nanotextured surfaces with hydrophobic patterns. *Energies* **2021**, *14*, 7543. [CrossRef]
27. Freitas, E.; Pontes, P.; Cautela, R.; Bahadur, V.; Miranda, J.; Ribeiro, A.P.C.; Souza, R.R.; Oliveira, J.D.; Copetti, J.B.; Lima, R.; et al. Pool boiling of nanofluids on biphilic burfaces: An experimental and numerical study. *Nanomaterials* **2021**, *11*, 125. [CrossRef]
28. Boinovich, L.B.; Emelyanenko, A.M. The behaviour of fluoro-and hydrocarbon surfactants used for fabrication of superhydrophobic coatings at solid/water interface. *Coll. Surf. A Phys. Eng. Asp.* **2015**, *481*, 167–175. [CrossRef]
29. Boinovich, L.B.; Emelyanenko, K.A.; Domantovsky, A.G.; Chulkova, E.V.; Shiryaev, A.A.; Emelyanenko, A.M. Pulsed laser induced triple layer copper oxide structure for durable polyfunctionality of superhydrophobic coatings. *Adv. Mater. Interfaces* **2018**, *5*, 1801099. [CrossRef]
30. Safonov, A.I.; Sulyaeva, V.S.; Gatapova, E.Y.; Starinskiy, S.V.; Timoshenko, N.I.; Kabov, O.A. Deposition features and wettability behavior of fluoropolymer coatings from hexafluoropropylene oxide activated by NiCr wire. *Thin Solid Films* **2018**, *653*, 165–172. [CrossRef]
31. Chinnov, E.A.; Shatskiy, E.N.; Khmel, S.Y.; Baranov, E.A.; Zamchiy, A.O.; Semionov, V.V.; Kabov, O.A. Enhancement of heat transfer at pool boiling on surfaces with silicon oxide nanowires. *J. Phys. Conf. Series* **2017**, *925*, 012033. [CrossRef]
32. Vladimirov, V.Y.; Khmel, S.Y.; Safonov, A.I.; Semionov, V.V.; Chinnov, E.A. Effect of fluoropolymer spots on pool boiling heat transfer. *J. Phys. Conf. Series* **2021**, *2119*, 012087. [CrossRef]
33. Može, M.; Zupančič, M.; Golobič, I. Investigation of the scatter in reported pool boiling CHF measurements including analysis of heat flux and measurement uncertainty evaluation methodology. *App. Therm. Eng.* **2020**, *169*, 114938. [CrossRef]
34. Rohsenow, W.M. *A Method of Correlating Heat Transfer Data for Surface Boiling of Liquids*; Technical Report No 5 (Heat Transfer Laboratory); MIT Division of Industrial Cooperation: Cambridge, MA, USA, 1951. Available online: https://dspace.mit.edu/handle/1721.1/61431 (accessed on 5 July 2022).

energies

MDPI

Article

Lattice Boltzmann Simulation of Optimal Biphilic Surface Configuration to Enhance Boiling Heat Transfer

Alexander V. Fedoseev [1,*], Mikhail V. Salnikov [1], Anastasiya E. Ostapchenko [1] and Anton S. Surtaev [1,2]

[1] Institute of Thermophysics SB RAS, Lavrentyev Ave. 1, Novosibirsk 630090, Russia
[2] Physics Department, Novosibirsk State University, Pirogova Str. 2, Novosibirsk 630090, Russia
* Correspondence: fedoseev@itp.nsc.ru

Abstract: To study the processes of boiling on a smooth surface with contrast wettability, a hybrid model was developed based on Lattice Boltzmann method and heat transfer equation. The model makes it possible to describe the phenomena of natural convection, nucleate boiling, and transition to film boiling, and, thus, to study heat transfer and the development of crisis phenomena in a wide range of surface superheats and surface wetting characteristics. To find the optimal configuration of the biphilic surface, at the first stage a numerical simulation was carried out for a single lyophobic zone on a lyophilic surface. The dependences of the bubble departure frequency and the departure diameter of the bubble on the width of the lyophobic zone were obtained, and its optimal size was determined. At the next stage, the boiling process on an extended surface was studied in the presence of several lyophobic zones of a given size with different distances between them. It is shown that in the region of moderate surface superheat, the intensity of heat transfer on biphilic surfaces can be several times (more than 4) higher compared to surfaces with homogeneous wettability. Based on numerical calculations, an optimal configuration of the biphilic surface with the ratios of the lyophobic zones' width of the order of 0.16 and the distance between the lyophobic zones in the range of 0.9–1.3 to the bubble departure diameter was found.

Keywords: pool boiling; heat transfer enhancement; biphilic surface; lattice Boltzmann method

Citation: Fedoseev, A.V.; Salnikov, M.V.; Ostapchenko, A.E.; Surtaev, A.S. Lattice Boltzmann Simulation of Optimal Biphilic Surface Configuration to Enhance Boiling Heat Transfer. *Energies* **2022**, *15*, 8204. https://doi.org/10.3390/en15218204

Academic Editor: Marco Marengo

Received: 12 October 2022
Accepted: 1 November 2022
Published: 3 November 2022

1. Introduction

To date, issues related to the enhancement of heat transfer during liquid boiling have received great attention from scientists and engineers, which is confirmed by numerous papers and reviews published in the past few years [1–5]. This is due to a wide range of practical applications in which the boiling process plays a key role. One of the most popular and effective methods for the enhancement of the heat transfer and increasing the critical heat flux (CHF) during boiling is the modification of the heat exchange surface. Surface modification can be implemented both by structuring the original surface using, for example, mechanical machining or laser texturing, and by fabrication of micro/nanostructured porous coatings [5]. As a result of heat exchange surface modifications, not only the morphology, but also the wetting characteristics can change. It is known that wetting properties have a significant effect on the heat transfer intensity and the development of crisis phenomena during liquid boiling. For example, the use of superhydrophilic micro- or nanoporous surfaces leads to a significant increase in the CHF value at parameters corresponding to water boiling (up to 2–3 times) [6,7]. In turn, the use of weakly hydrophobic coatings leads to a decrease in the temperature of the onset of nucleate boiling (ONB), to an increase in the nucleation site density and the heat transfer enhancement in the regions of low heat fluxes [8–11]. However, the CHF significantly decreases when using hydrophobic and superhydrophobic coatings [11–13]. In particular, in [10,11] during water boiling on flat and tubular hydrophobic heat exchange surfaces (contact angle $\theta \approx 125°$), heat transfer enhancement was found in the region of low heat fluxes (q < 100–150 kW/m^2) compared to heat transfer on the base hydrophilic surface. Moreover, the use of the developed hydrophobic coatings can significantly reduce the temperature of the onset of water boiling (up to 2–3° K), especially during boiling at subatmospheric pressures [11]. However, the CHF decreases significantly when using hydrophobic and superhydrophobic coatings [11–13]. For example, in an experimental study [12] it was shown that the usage of a superhydrophobic heated surface (contact angle $\theta \approx 155°$), vapor film is almost immediately formed at 1–2° K of superheating, covering the entire surface, thus the heat transfer is reduced just after the onset of boiling.

One of the modern trends in the field of material science in relation to the problems of improvement of pool boiling performance is the fabrication of biphilic surfaces or, in other words, surfaces with contrast wettability. The use of biphilic surfaces, as shown by the analysis of recent papers [14–17], permits simultaneously increasing the heat transfer intensity by reducing the bubbles activation threshold and increasing CHF due to more efficient wetting of dry spots in precrisis modes and modulation of the two-phase flows near the heated wall. In particular, it was shown in experiments [14,15] that the use of biphilic coatings can lead to an increase in the heat transfer coefficient and CHF values up to 5 and 2 times, respectively. At the same time, the task of determination of the biphilic surface configuration for the optimal boiling conditions remains open. These configurations vary for different liquids and boiling regimes including pressure change.

The search for the optimal configuration of a surface with contrast wettability based on experimental studies is a rather laborious task, the complexity of which is primarily determined by the creation of stable coatings with the ability to control the geometric parameters of the structure at various scales, including micro and nanoscales. An additional complication is the requirement for high-precision high-speed measurements to study the evolution of vapor bubbles and a two-phase layer near the surface using a set of modern experimental methods. With the rapid development of computer technology, computational fluid dynamics (CFD) methods have proven to be a good alternative to experiment for studying two-phase flows, including those in which phase transitions occur [18–26]. The classical approach to modeling two-phase flows, including boiling, is reduced to the numerical solution of the Navier–Stokes equations in combination with an additional method for tracking the two-phase interface. These methods include the level set method [19–22], the volume of liquid (VOF) method [23,24], or the combined VOSET method [25,26]. A common problem with these methods is that they do not allow modeling of the nucleation process, which should be considered at the micro or nano scale. This process is crucial for modeling the life cycle of vapor bubbles, in particular, the stage of waiting for the appearance of nuclei, the temperature of the onset of boiling, and the density of nucleation sites. Without the ability to model these characteristics, it is impossible to calculate the boiling curves for an ensemble of bubbles that form on extended surfaces. To simulate micro-nanoscale phenomena, such as the process of nucleation, the molecular dynamics method is well applicable [27–29]. However, its application on a macro scale is difficult due to usually limited computational power.

Lattice Boltzmann Method (LBM) implies a different approach to multiphase flows simulation. In this method, the medium is represented in the form of ensembles of pseudoparticles, for which the kinetic equation is solved in the discrete formulation. Only a small set of pseudoparticle velocities is possible, such that the velocities are directed towards neighboring nodes of the spatial lattice. This method belongs to the mesoscopic category, and occupies an intermediate position between the microscopic method of molecular dynamics and the macroscopic CFD approaches. Despite the complexity of the Boltzmann kinetic equation in the classical formulation, its discrete counterpart is simple and ideally suited for parallel computing. Over 30 years of LBM development, it has proven itself well for modeling both single-phase and multi-phase flows [30,31]. In [32], the LBM with a pseudopotential was presented, in which the interaction function between pseudoparticles is specified. With this approach, phase separation is obtained in a natural way as a consequence of the forces of interaction of pseudoparticles at the lattice sites. The action of body forces can be specified in different ways, of which the Guo method [33], He method [34] and the exact difference method (EDM) [35] are the most effective. Another important advantage of LBM is the way of specifying the condition at the fluid-solid boundary, the so-called bounce-back condition [36]. This makes it possible to simply set a solid boundary of an arbitrary shape [36], as well as to set an arbitrary local value of the wetting angle [37].

The advantages stated above made it possible to use LBM to model and study the pool boiling processes. Using this method, it is possible to simulate the process of vapor phase nucleation without the need to specify additional initial conditions obtained empirically. The authors of [38,39] used LBM to simulate the processes of growth and detachment of single bubbles during boiling. The dependence of the departure diameter and the frequency of bubble departure on the wetting contact angle and the Jacob number was studied. In [40–44], boiling curves were calculated for smooth surfaces with different uniform wettability. It was shown that the boiling crisis for a lyophobic heater occurs at lower superheats than for a lyophilic one, and an increase in the wetting contact angle leads to a decrease in the CHF. Recently, works have been presented on the LBM simulation of heat transfer processes during boiling on surfaces with contrast wettability [45,46]. The papers simulated boiling on a lyophilic surface with the inclusion of lyophobic zones. The authors demonstrated the possibility of increasing the heat flux compared to a homogeneous lyophilic surface.

The purpose of the paper is to develop a hybrid model based on LBM and heat transfer equations capable of simulating the process of boiling on surfaces with contrast wettability in a wide range of surface superheats up to the boiling crisis development. The model will permit obtaining the boiling curves for heat exchange surfaces with homogeneous and heterogeneous wettability, to study the effect of the size of lyophobic zones on a lyophilic base surface and of the distance between them, and to determine the optimal biphilic surface configuration for efficient heat transfer.

2. Model

For the study of pool boiling processes, a hybrid model based on the standard pseudopotential multiphase Lattice Boltzmann method coupled with the heat transfer equation is used. Below a brief overview of the model is presented while a more detailed description can be found in [39,41]. The basic equation for the time evolution of the density distribution function f_i in a discrete form can be presented in the form:

$$f_i(\vec{x} + \vec{c}_i\Delta t, t + \Delta t) = f_i(\vec{x}, t) + \Omega_i(\vec{x}, t) + S_i(\vec{x}, t), \tag{1}$$

where $\vec{x}$ is the position and t is the dimensionless time, $\vec{c}_i$ are the unit lattice vectors, Ω_i is the collision operator, which describes the particles' interaction, and S_i is the change of the distribution function due to the action of the bulk forces. In the paper, a two dimensional D2Q9 lattice with nine directions is used: $|\vec{c}_i|$ = (0, 1, $\sqrt{2}$), i = {1, 2, ..., 5, 6, ..., 9}. The density ρ and the velocity $\vec{u}$ of the medium in each point can be obtained from the moments of the distribution function:

$$\rho(\vec{x}, t) = \sum_i f_i(\vec{x}, t), \ \rho\vec{u}(\vec{x}, t) = \sum_i \vec{c}_i f_i(\vec{x}, t). \tag{2}$$

The Bhatnagar–Gross–Krook (BGK) approximation for a collision operator is used:

$$\Omega_i(\vec{x}, t) = -\frac{f_i - f_i^{eq}}{\tau}. \tag{3}$$

where f_i^{eq} is the equilibrium distribution function, τ is the relaxation time connected with the kinematic viscosity ν as $\tau = 0.5 - \nu/c_s^2$, and where $c_s{}^2 = (1/3)\Delta x^2/\Delta t^2$ is the lattice speed of sound; Δx and Δt are the spatial and time steps.

For the body force term, the exact difference method [35] is used:

$$S_i = f_i^{eq}(\rho, \vec{u} + \Delta\vec{u}) - f_i^{eq}(\rho, \vec{u}) \tag{4}$$

where the change of velocity $\vec{u}$ is determined by the force acting on the node: $\Delta\vec{u} = \vec{F}/\rho$. The total force $\vec{F}$ acting on the particles consists of the following components:

$$\vec{F}(\vec{x}) = \vec{F}_{\text{int}}(\vec{x}) + \vec{F}_s(\vec{x}) + \vec{F}_g(\vec{x}), \tag{5}$$

where $\vec{F}_{\text{int}}(\vec{x})$ is the particles' interaction force responsible for phase separation, $\vec{F}_s(\vec{x})$ describes the interaction of the medium with the solid surface, and $\vec{F}_g(\vec{x}) = g(\rho - \rho_{avg})$ is the gravity force, where ρ_{avg} is the average density of the liquid/vapor medium; g is free fall acceleration.

The equation of state (EOS) of the simulated fluid is defined by $\vec{F}_{\text{int}}(\vec{x})$ written as a gradient of a pseudopotential Ψ as proposed in [32]. The gradient of the pseudopotential is approximated with the finite-difference scheme [35]. In the paper, the Peng–Robinson equation of state [47] is used to determine the pressure dependence $P(\rho,T)$ on the density ρ and the temperature T of the liquid/vapor medium.

In the absence of the fluid-solid interaction (adhesive force), a liquid-solid contact angle θ is equal to 90°. To adjust the value of the contact angle, an approach proposed in [37] is used, and an extra force $\vec{F}_s(\vec{x})$ is added to the nodes near the solid boundary. The contact angle is modeled by varying the strength and direction of the force $\vec{F}_s(\vec{x})$ as described in detail in [37,41]. In the present paper, the contact angles θ = 38°, 90° and 116° are used for the simulation of the lyophilic, neutral and lyophobic surfaces.

Evolution of the heat transfer in the computational region is described by the heat conduction equation taking into account diffusion, convection, work of pressure forces, as well as phase transition [41,42]. In the model, the dimensionless parameters in the lattice units are used. The values of pressure P, temperature T and density ρ are expressed in units P_c, T_c and ρ_c, which correspond to the parameters of the fluid at the critical point. In dimensionless form this equation reads:

$$\frac{\partial T}{\partial t} + \vec{u}_f \vec{\nabla} T = \frac{1}{\rho_c \rho c_v} \vec{\nabla}\left(\lambda \vec{\nabla} T\right) \frac{\Delta t}{\Delta x^2} - \frac{T}{\rho_c \rho c_v} \frac{P_c}{T_c} \left(\frac{\partial p_{EOS}}{\partial T}\right)_\rho \vec{\nabla} \cdot \vec{u}_f, \tag{6}$$

where c_v and λ are the heat capacity and thermal conductivity coefficient. Here, $\vec{u}_f = \vec{u} + \Delta\vec{u}/2$ is the physical velocity of the continuum. The last term in the equation implicitly accounts for the latent heat of vaporization, which is the main source of enhancement of the heat flux under the onset of nucleate boiling.

The computational domain consists of a liquid/vapor medium and a metal heater (see Figure 1). In the paper, 800 spatial cells in a horizontal direction (axis x) and 500 spatial cells in a vertical direction (axis y) are used. The metal heater with a thickness of n_h = 30 spatial cells is placed on the bottom boundary. Spatial and time steps are equal to $\Delta x = 20 \times 10^{-6}$ m and $\Delta t = 2.5 \times 10^{-6}$ s, respectively. Periodic boundary conditions are applied to the left and the right boundaries. On the top boundary, a constant temperature $T_0 = 0.9\ T_c$ and corresponding pressure P_0 are specified. Inside the metal heater, only heat diffusion Equation (6) is solved, with constant temperature T_h condition imposed at the bottom. The calculations were performed for different values of superheat ΔT, i.e., the temperature difference between the top and the bottom boundaries of the solution region, $\Delta T = T_h - T_0$. Thermal conductivity and heat capacity of the metal heater are set to λ_h = 20 W/m/K and $c_h = 3 \times 10^6$ J/m^3, respectively. The thermodynamic properties of the fluid, e.g., heat capacity, thermal conductivity, and viscosity are evaluated based on the current density of the liquid/vapor medium and are parameterized based on the properties of the liquid and its vapor.

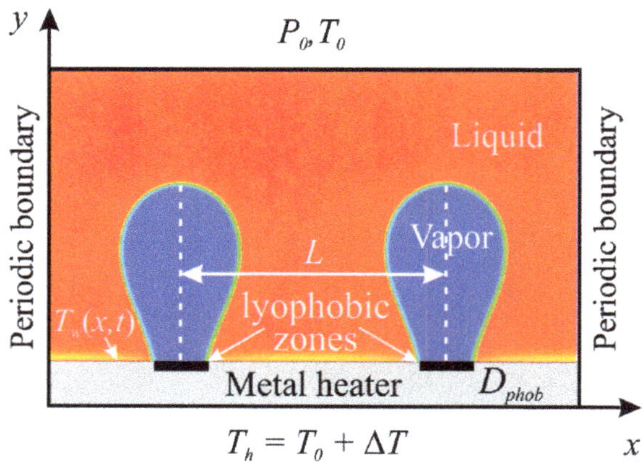

Figure 1. Computational domain and boundary conditions. Layout of the lyophobic zones on the lyophilic surface: D_{phob} is the width of the lyophobic zones, L is the distance between the lyophobic zones.

The phase diagram for some instant at boiling on a biphilic surface is also presented in Figure 1. Areas with the gas phase are shown in blue while the liquid phase is shown in red. The distance between the neighbor lyophobic zones is L and the width of each lyophobic zone is D_{phob}.

3. Results

3.1. Boiling on the Surfaces with Homogenous Wettability

At the initial stage, the boiling curves $\langle q\rangle(\Delta T')$ were calculated for the smooth surfaces with homogeneous wettability including lyophobic and lyophilic states, where $\Delta T' = \langle T_w\rangle - T_{sat}$ is the surface superheat and T_{sat} is the saturation temperature (Figure 2). Average heater surface temperature $\langle T_w\rangle$ is obtained by averaging the temperature $T_w(x,t)$ over time and the heater surface. In calculations, to obtain each point of the dependence $q(\Delta T')$, temperature value is set on the lower wall of the heater T_h = const. The heater surface temperature $T_w(x,t)$ depends on the heat transfer processes and evolves according to the Equation (6), see [41]. In turn, the time and the surface averaged heat flux $\langle q\rangle$ through the metal heater of the height $H_h = n_h\, h$ is determined as $\langle q\rangle = -\lambda_h\,(\langle T_w\rangle - T_h)/H_h$.

As can be seen from the data presented in Figure 2, at low superheats, the boiling curves $q(\Delta T')$ do not depend on the value of the wetting angle, since the heat transfer in this region occurs in the mode of single-phase natural convection. After the onset of boiling, the slope of the $q(\Delta T')$ curves sharply increases, which corresponds to an increase in the intensity of heat transfer with the transition from natural convection to nucleate boiling. It can be seen that the temperature corresponding to ONB (onset of nucleate boiling) noticeably decreases with increasing contact angle, which is in qualitative agreement with the results of experimental studies [8–13]. The analysis of the curves in Figure 2 shows that the heat transfer depends in a complex way on the surface superheats and surface wettability. For example, in the region of low surface superheat, the removal heat flux increases with increasing contact angle, which qualitatively agrees with the results of experiments [9–11], in which it was shown that for weakly hydrophobic surfaces (θ = 116–130°) at low input heat fluxes, an enhancement of heat transfer is observed in comparison with hydrophilic samples. However, as calculations show, the situation changes dramatically in the area of high wall superheats. It can be seen that the value of the CHF decreases significantly with an increase in the contact angle, which also qualitatively agrees with the experimental results [11–13]. Moreover, the boiling crisis on the lyophobic surface can be observed even before the onset of nucleate boiling on the lyophilic surface at a given wall superheat.

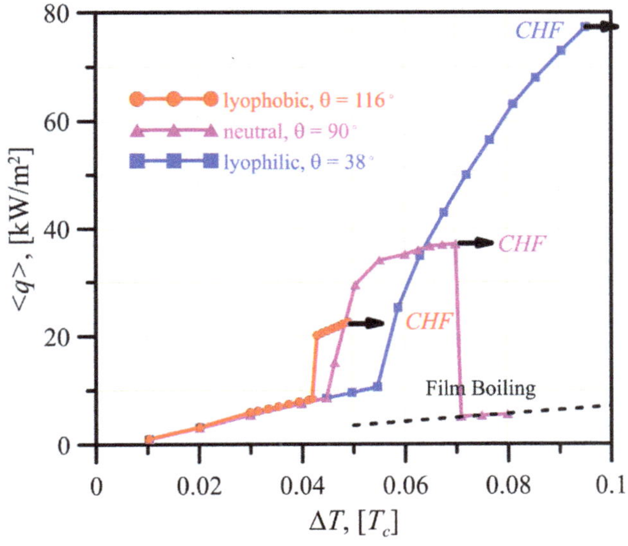

Figure 2. Boiling curves $q(\Delta T')$ for the smooth surfaces with homogeneous wettability: lyophilic surface (θ = 38°), neutral surface (θ = 90°), lyophobic surface (θ = 116°).

Figure 3 shows density contour plots of boiling at low wall superheat (ΔT = 0.045 T_c) on surfaces with different contact angles. As can be seen from the pictures, at low surface superheats the boiling on homogeneous lyophobic surface (θ = 116°) resembles a transitional boiling regime, when part of the surface is occupied by vapor phase with local sites of film boiling. In general, this picture is similar to the results of experimental studies, in which it was shown that boiling on weakly hydrophobic surfaces in the region of low heat fluxes is characterized by the presence of large-scale sessile bubbles with periodic separation of the vapor phase from liquid-vapor interface [10–12]. It has also been shown in the vast majority of experiments [10–12,48,49] that for lyophobic surfaces, the angle of inclination of the liquid meniscus with respect to the solid body exceeds 90°, and the vapor bubble detaches at some distance from the heater surface. This leads to the fact that after the departure of the vapor bubble, part of the vapor always remains on the surface, which leads to the degeneration of the waiting stage of the appearance of a bubble in the life cycle of the nucleation site. Indeed, the results of LBM simulation on lyophobic surfaces ($\theta \geq 90°$) demonstrate a qualitatively similar behavior of the triple contact line during evolution of vapor bubbles on the heating surface. At the same wall superheat, as can be seen form Figure 3c, the heat transfer on the lyophilic surface (θ = 38°) occurs in the free convection mode without activation of nucleation sites.

Figure 3. The density contour plots illustrating the boiling process at low surface superheat, $\Delta T = 0.045\ T_c$. (**a**) lyophobic surface (θ = 116°), (**b**) neutral surface (θ = 90°) (**c**) lyophilic surface (θ = 38°).

With an increase in wall superheat $\Delta T = 0.06\ T_c$, the boiling behavior on surfaces with different wettability changes (Figure 4). As can be seen from Figure 4a, in this case for lyophobic surface, the stable film boiling mode is observed, and the entire surface of the heater is covered with a vapor phase with periodic detachment of vapor bubbles from the film surface. For a neutral surface (Figure 4b), with an increase in the input heat flux, several nucleation sites can simultaneously be activated on the surface, which also leads to an increase of vaporization rate compared to the case of low superheat. However, large bubbles with the area bounded by the triple contact line larger than the bubble departure diameter may also appear. For a lyophilic surface at given superheat (Figure 4c), vapor bubbles are already activated in comparison with the case shown in Figure 3c. However, the boiling process seems to have a periodic character, when several nucleation sites are activated at once, and after the detachment of vapor bubbles, a long waiting stage is observed. This leads to noticeable pulsations in the space averaged wall temperature over time. This character of boiling can be related to the fact that an absolutely smooth surface is specified in the simulation model, when there are no cavities which are potential sites for the activation of vapor bubbles. At the same time, the evolution of individual bubbles qualitatively corresponds to experimental observations of the dynamics of growth and detachment of bubbles during boiling of a wide class of liquids on lyophilic surfaces [19–22,38,39].

Figure 4. The density contour plots illustrating the boiling process at moderate surface superheat, $\Delta T = 0.06\ T_c$. (**a**) lyophobic surface (θ = 116°), (**b**) neutral surface (θ = 90°) (**c**) lyophilic surface (θ = 38°).

3.2. Bubble Dynamic at Boiling above a Single Lyophobic Zone

In our opinion, the choice of the configuration of the biphilic pattern surface is determined primarily by the selection of the optimal sizes of the lyophobic zones and the distances between them on the lyophilic base. Therefore, at the first stage, the process of vapor bubble growth and detachment for a single lyophobic zone with different sizes (contact angle θ = 116°) located in the center of the smooth lyophilic surface of the heater (θ = 38°) was simulated. In LBM simulation, the following range of widths of the lyophobic zone D_{phob} = 0.3–3 mm was considered, which is noticeably larger than the critical radius of vapor nuclei and smaller or comparable with the bubble departure diameter. Moreover, this range of sizes of the lyophobic zone is, as a rule, used to create biphilic surfaces in experimental works [14–17].

Figure 5 shows the characteristic temporal evolution of a single bubble for a given width of the lyophobic zone D_{phob} = 1.5 mm and a heater temperature $T_h = 0.95\ T_c$. The size of the computational domain was 400 cells horizontally (12 mm) and 600 cells vertically (18 mm). In general, the observed picture of the process is similar to boiling on a lyophobic surface; however, there is some limitation on the size of the contact line region associated with the finite size of the lyophobic spot.

Then, various key characteristics of bubble evolution were studied and dependences of the departure diameter D_d and the departure frequency ν_d of bubbles on the width of the lyophobic zone D_{phob} were plotted (Figure 6). It can be seen that the bubble departure diameter D_d increases almost linearly with the width of the lyophobic zone D_{phob} and changes by 25% in the studied range (Figure 6a). At the same time, on the dependence of the bubble departure frequency ν_d on D_{phob}, shown in Figure 6b, a maximum is observed at D_{phob} = 450 μm. The higher the bubble departure frequency, the more energy is spent on vaporization averaged over time. The thermal analysis of calculations results also demonstrate that the maximum heat flux removed from the surface is observed also for a lyophobic zone with a width of D_{phob} = 450 μm. Therefore, further we will consider this width of the lyophobic zone as the optimal.

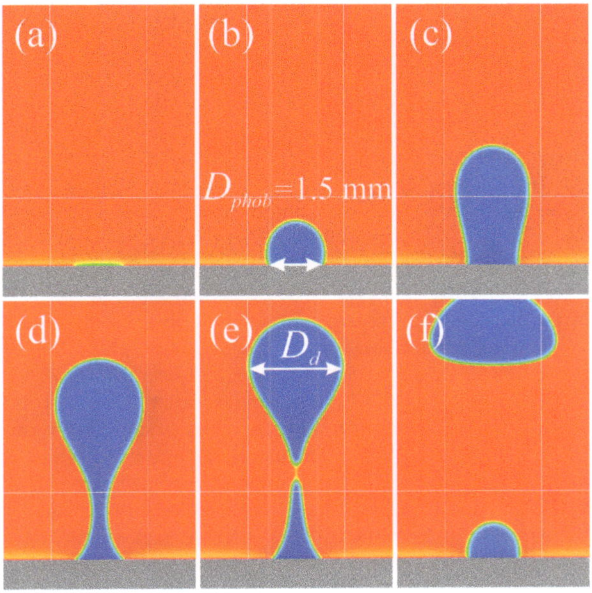

Figure 5. Temporal evolution of the growth and departure of a single bubble: (**a**) t = 0.15 s, (**b**) t = 0.175 s, (**c**) t = 0.375 s, (**d**) t = 0.475 s, (**e**) t = 0.5 s, (**f**) t = 0.525 s. Lyophobic zone size D_{phob} = 1.5 mm, horizontal region size is 12 mm, heater temperature T_h = 0.95 T_c.

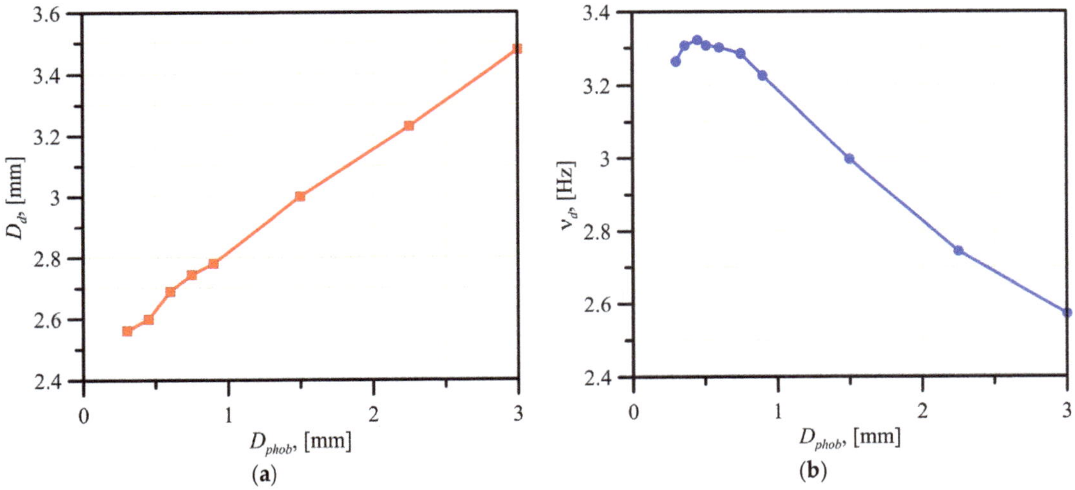

Figure 6. (**a**) Bubble departure diameter D_d and (**b**) departure frequency ν_d depending on the lyophobic zone width D_{phob}. Heater temperature T_h = 0.95 T_c.

3.3. Boiling on the Surfaces with Contrast Wettability

At the next stage, the boiling process was simulated on the surfaces with contrast wettability by changing the distance L between the lyophobic zones (i.e., pitch size) with an optimal spot width D_{phob} = 450 μm. The size of the computational domain was 800 cells horizontally (24 mm) and 500 cells vertically (15 mm). Simulation was carried out for different numbers N = 3, 5, 7, 10, 15 and 20 of lyophobic zones equidistantly located on the lyophilic surface, which corresponded to the range of the distances between the spots L = 1.2–12 mm, and the ratio L/D_d = 0.46–4.61. Thus, such characteristic cases were considered when the distance between the centers of bubble nucleation is less $L/D_d < 1$, corresponds to $L/D_d \sim 1$, and is greater $L/D_d > 1$ than the departure diameter of the bubble. An additional analysis of the density contour plots and temperature fields was carried out in a wide range of input heat flux, which permitted determination of the cause-and-effect relationships between the dependence of the heat transfer efficiency and the boiling behavior for the various configurations of the patterned biphilic surface.

The analysis of boiling curves presented in Figure 7 shows that in the regime of single phase convection ($\Delta T \leq 0.044\ T_c$) the dependencies for surfaces with contrast wettability coincide with the surface with homogeneous wettability. The onset of nucleate boiling on the surfaces with contrast wettability is observed at the same temperature of the heater as on the pure lyophobic surface ($T_{ONB,phob} \approx 0.944\ T_c$), but much lower than on the lyophilic surface ($T_{ONB,phil} \approx 0.955\ T_c$). The calculated curves show that with the development of boiling at $\Delta T > 0.044\ T_c$, the removed heat flux $<q>$ from the biphilic surfaces increases significantly compared to surfaces with homogeneous wettability at a given wall superheat. Calculations also show that at certain surface superheats ($\Delta T \approx 0.07\ T_c$), the intensity of heat transfer on biphilic surfaces begins to decrease in relation to boiling heat transfer on lyophilic surface. It is also seen that at high surface superheat $\Delta T > 0.07\ T_c$, the removed heat flux $<q>$ from the surfaces with contrast wettability decreases with a decrease of the distance L between the lyophobic zones that corresponds to larger number of the zones. It should be noted that the CHF on the surfaces with contrast wettability is lower than the CHF on the lyophilic surface, and the value of the CHF decreases with a decrease of the distance between the lyophobic zones.

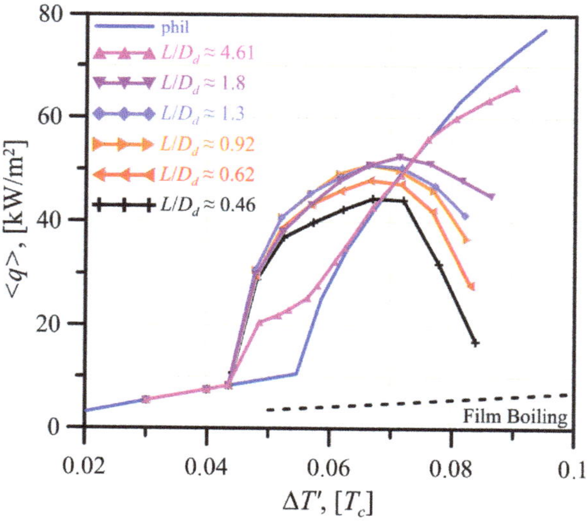

Figure 7. Boiling curves $q(\Delta T')$ for the smooth surfaces with contrast wettability. Lyophobic zone width D_{phob} = 450 μm, space separation between the zones L/D_d = 0.46, 0.62, 0.92, 1.3, 1.8 and 4.61.

Figure 8 shows the density contour plots of boiling on surfaces with contrast wettability for various numbers of lyophobic zones on the lyophilic substrate, $N = 2$, 7 and 20, at moderate surface superheat $\Delta T = 0.065\ T_c$. On a surface with $N = 2$ lyophobic zones, the formation of bubbles occurs mainly above these zones, and after the bubble departure, the vapor phase always remains above them. The bubble base then may extend beyond the boundaries of the lyophobic zone due to the small width of the zone itself. The distance between the zones for $N = 2$ exceeds the separation diameter of the bubble by several times, $L/D_d = 4.61$, and therefore the bubbles do not interact with each other. For a surface with $N = 7$ lyophobic zones, the distance between lyophobic zones becomes comparable to the bubble departure diameter, $L/D_d = 1.3$. Neighbor bubbles begin to interact with each other. As a result of the merging of two adjacent vapor bubbles, some part of the heating surface can be covered by a vapor phase. The usage of the surface with large number of lyophobic zones, $N = 20$, and, accordingly, a small distance between the spots, $L/D_d = 0.62$, leads to coalescence of bubbles from neighboring lyophobic zones in the early stages of their evolution, which ultimately leads to an increase in the area occupied by dry spots and a decrease in the intensity of heat transfer during boiling compared to other configurations of biphilic surfaces described above. Figure 7 shows that at moderate surface superheat $\Delta T = 0.065\ T_c$, the heat flux removed from the surface with large distance between the lyophobic zones, $L/D_d = 4.61$, is much lower than from the surfaces with $L/D_d = 1.3$ and $L/D_d = 0.62$. However, the heat flux removed from the surface with low distance between the zones $L/D_d = 0.62$ is less than from a surface with $L/D_d = 1.3$. At high surface superheat $\Delta T > 0.07\ T_c$, a significant part of the heating surface is covered by a vapor film due to merging of several bubbles. These local areas of film boiling are quite stable over time, which leads to a noticeable decrease in the intensity of heat transfer for biphilic surfaces with $L/D_d < 1.8$.

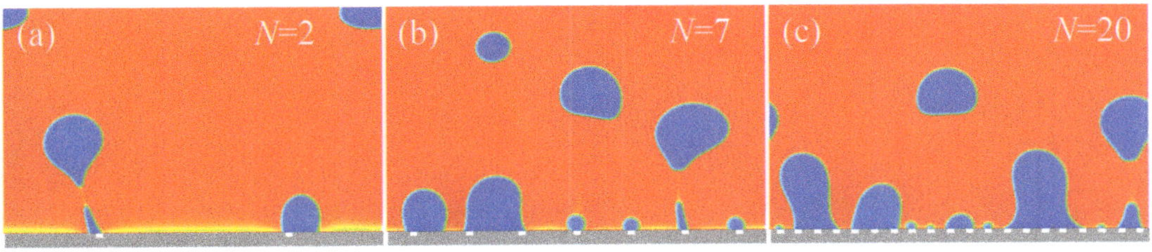

Figure 8. The density contour plots of boiling at $\Delta T = 0.065\ T_c$ for different numbers of the lyophobic zones: (**a**) $N = 2$ ($L/D_d \approx 4.61$), (**b**) $N = 7$ ($L/D_d \approx 1.3$), (**c**) $N = 20$ ($L/D_d \approx 0.62$).

It should be noted that the degradation of heat transfer on modified surfaces with contrast properties compared to the unmodified lyophilic surface was also observed experimentally. In [50], degradation of the heat transfer has been observed with increasing the size of hydrophobic spots. In [51], the formation of a "gigantic single bubble" on the pattered surface was observed as a result of bubbles merging at high surface superheats. A similar effect of heat transfer degradation was observed on a structured surface at a small distance between cavities [52].

3.4. Discussion

Finally, we tried to generalize the obtained results on the enhancement of heat transfer at pool boiling on surfaces with contrast and homogeneous wettability. Let us consider the ratio of the heat transfer coefficient on a biphilic surface to the heat transfer coefficient on a lyophilic surface, HTC/HTC_{phil}. Figure 9 shows the dependences of enhancement factor HTC/HTC_{phil} on the distance between the lyophobic zones for various values of the heater temperature T_h. It can be seen that for moderate heater temperatures, $T_h < 0.96\ T_c$, the enhancement factor on surfaces with contrast wettability is greater than unity. However, at high heater temperatures, $T_h > 0.97\ T_c$, the enhancement factor on the modified surfaces decreases due to the presence of wide unwetted areas. It was also found that the dependences of the enhancement factor on the distance between zones L/D_d have a maximum around $L/D_d \sim 1.1$. The maximum values of the enhancement factor $HTC/HTC_{phil} \approx 4$ at heater temperature $T_h = 0.955\ T_c$ were obtained for surfaces with parameters $L/D_d \approx 0.92$ and 1.3. Thus, as a result of the performed numerical simulations, it can be concluded that for moderate surface superheat the optimal distance between the lyophobic zones should be of the order of the bubble departure diameter, $0.9 \leq L/D_d \leq 1.3$ and the ratio of the width of the lyophobic zone to the bubble departure diameter should be of the order of $D_{phob}/D_d \sim 0.16$. Such distance between the lyophobic zones relative to the bubble departure diameter is consistent with the results of experiments [53], in which heat transfer and bubble dynamics were simultaneously analyzed on biphilic surfaces with different configurations.

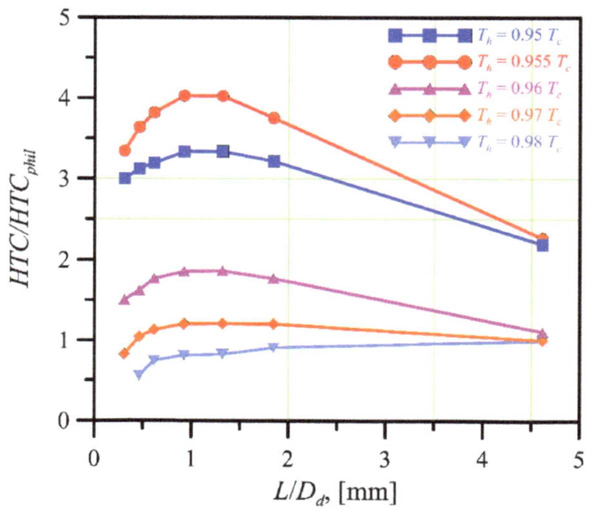

Figure 9. Enhancement factor for heat transfer on the surfaces with contrast wettability depending on the distance between the lyophobic zones L/D_d for different heater temperatures T_h.

It should be emphasized that the simulation results obtained in the paper with the help of hybrid LBM model for the boiling on ideally smooth surfaces do not take into account a number of important aspects related to real experimental situations. First of all, the roughness of heat exchange surfaces leads to some stabilization of the bubbles and influences the nucleation process. Although the liquid parameters given in the model are taken as close as possible to water physical properties, the obtained results correspond to some kind of a model fluid. In the future, it is necessary to carry out simulations varying the properties of the fluid in order to search for universal regularities. It should be also noted that in the model a low liquid to vapor densities ratio ($\rho_l/\rho_v \sim 25$) is considered, that corresponds to the case close to the critical point. Furthermore, 3D modeling could lead to a better agreement with available experimental data than obtained in this work with 2D simulations. Nevertheless, the presented results qualitatively describe the main characteristics of the boiling process, and the abovementioned shortcomings set a task for improvement of the model in the future.

4. Conclusions

In this study, to simulate the pool boiling on smooth surfaces with contrast wettability, a hybrid model based on the lattice Boltzmann method and the heat transfer equation in a two-phase medium was used. As a result of numerical simulation, boiling curves were calculated for surfaces with uniform wettability (lyophilic, neutral and lyophobic surfaces). The process of boiling over a single lyophobic zone on a lyophilic surface was studied, and dependences of the bubble departure frequency and bubble departure diameter on the width of the spot were obtained that permit calculation of its optimal size. The pool boiling on the surfaces with different configuration of the patterns was studied and the boiling curves were calculated.

The following conclusions were drawn:

- The calculations showed that the use of the surfaces with contrast wettability permits a substantial decrease in the onset of nucleate boiling compared to a bare lyophilic surface. The ONB of biphilic surfaces occurs approximately at the same superheats as on a homogeneous lyophobic surface.
- Based on the simulations of a bubble dynamic at boiling above a single lyophobic zone and on analysis of key characteristics of bubble evolution, i.e., the bubble departure diameter D_d and the bubble departure frequency ν_d, the optimal width of the lyophobic spot $D_{phob}/D_d \sim 0.16$ was obtained.
- It was shown that at moderate surface superheat, the heat transfer at boiling on surfaces with contrast wettability is significantly higher than the heat transfer on the surfaces with uniform wettability.
- In terms of heat transfer performance, the optimal configuration of the biphilic surface at moderate surface superheats was determined: the ratio of the width of the lyophobic zone $D_{phob}/D_d \sim 0.16$ and the distance between the lyophobic zones $0.9 \leq L/D_d \leq 1.3$ to the bubble departure diameter permits enhancement of the heat transfer by more than 4 times. The presented comprehensive studies made it possible to better understand the physics of the boiling process on surfaces with contrast wettability, and to identify the mechanisms of heat transfer enhancement/degradation. Based on the obtained simulation results, an optimal configuration of lyophobic zones on a lyophilic heating surface is proposed for a significant enhancement of boiling heat transfer. This information can already be used for future experiments to simplify the search for optimal configurations of biphilic surfaces for enhancement of boiling heat transfer. At the same time, it is necessary to further improve the numerical model, which would allow not only to control the surface wetting properties, but also to set morphological parameters, as well as to vary the properties of the liquid to simulate real coolants, heat exchange devices and thermal stabilization systems operating in various boiling conditions.

Author Contributions: Conceptualization, A.V.F. and A.S.S.; methodology, A.V.F. and A.E.O.; formal analysis, A.V.F. and A.S.S.; writing—original draft preparation, A.V.F., M.V.S., A.E.O. and A.S.S.; writing—review and editing, A.V.F., M.V.S., A.E.O. and A.S.S. All authors have read and agreed to the published version of the manuscript.

Funding: This research was funded by Russian Science Foundation grant number 22-29-01251.

Data Availability Statement: Data are available from the authors.

Conflicts of Interest: The authors declare no conflict of interest.

References

1. Liang, G.; Mudawar, I. Review of pool boiling enhancement by surface modification. *Int. J. Heat Mass Transf.* **2019**, *128*, 892–933. [CrossRef]
2. Sajjad, U.; Sadeghianjahromi, A.; Muhammad, H.A.; Wang, C.C. Enhanced pool boiling of dielectric and highly wetting liquids—A review on enhancement mechanisms. *Int. J. Heat Mass Transf.* **2020**, *119*, 104950. [CrossRef]
3. Mehralizadeh, A.; Shabanian, S.; Bakeri, G. Effect of modified surfaces on bubble dynamics and pool boiling heat transfer enhancement: A review. *Therm. Sci. Eng. Prog.* **2019**, *15*, 100451. [CrossRef]
4. Khan, S.A.; Atieh, M.A.; Koç, M. Micro-nano scale surface coating for nucleate boiling heat transfer: A critical review. *Energies* **2018**, *11*, 3189. [CrossRef]
5. Surtaev, A.; Serdyukov, V.; Pavlenko, A. Nanotechnologies for thermophysics: Heat transfer and crisis phenomena at boiling. *Nanotech. Russ.* **2016**, *11*, 696–715. [CrossRef]
6. Rahman, M.; Olceroglu, E.; McCarthy, M. Role of wickability on the critical heat flux of structured superhydrophilic surfaces. *Langmuir* **2014**, *30*, 11225–11234. [CrossRef]

7. Xie, S.; Beni, M.S.; Cal, J.; Zhao, J. Review of critical-heat-flux enhancement methods. *Int. J. Heat Mass Transf.* **2018**, *122*, 275–289. [CrossRef]
8. Bourdon, B.; Rioboo, R.; Marengo, M.; Gosselin, E.; De Coninck, J. Influence of the wettability on the boiling onset. *Langmuir* **2012**, *28*, 1618–1624. [CrossRef]
9. Betz, A.; Jenkins, J.; Kim, C.; Attinger, D. Boiling heat transfer on superhydrophilic, superhydrophobic, and superbiphilic surfaces. *Int. J. Heat Mass Transf.* **2013**, *57*, 733–741. [CrossRef]
10. Safonov, A.I.; Kuznetsov, D.V.; Surtaev, A.S. Fabrication of Hydrophobic Coated Tubes for Boiling Heat Transfer Enhancement. *Heat Transf. Eng.* **2020**, *42*, 1390–1403. [CrossRef]
11. Surtaev, A.; Koşar, A.; Serdyukov, V.; Malakhov, I. Boiling at subatmospheric pressures on hydrophobic surface: Bubble dynamics and heat transfer. *Int. J. Therm. Sci.* **2022**, *173*, 107423. [CrossRef]
12. Teodori, E.; Valente, T.; Malavasi, I.; Moita, A.S.; Marengo, M.; Moreira, A.L.N. Effect of extreme wetting scenarios on pool boiling conditions. *Appl. Therm. Eng.* **2017**, *115*, 1424–1437.
13. Phan, H.; Caney, N.; Marty, P.; Colasson, S.; Gavillet, J. Surface wettability control by nanocoating: The effects on pool boiling heat transfer and nucleation mechanism. *Int. J. Heat Mass Transf.* **2009**, *52*, 5459–5471. [CrossRef]
14. Motezakker, A.; Sadaghiani, A.K.; Çelik, S.; Larsen, T.; Villanueva, L.G.; Kosar, A. Optimum ratio of hydrophobic to hydrophilic areas of biphilic surfaces in thermal fluid systems involving boiling. *Int. J. Heat Mass Transf.* **2019**, *135*, 164–174.
15. Može, M.; Zupančič, M.; Golobič, I. Pattern geometry optimization on superbiphilic aluminum surfaces for enhanced pool boiling heat transfer. *Int. J. Heat Mass Transf.* **2020**, *161*, 120265.
16. Cheng, H.C.; Jiang, Z.X.; Chang, P.H.; Chen, P.H. Effects of difference in wettability level of biphilic patterns on copper tubes in pool boiling heat transfer. *Exp. Therm. Fluid Sc.* **2021**, *120*, 110241.
17. Serdyukov, V.; Patrin, G.; Malakhov, I.; Surtaev, A. Biphilic surface to improve and stabilize pool boiling in vacuum. *Appl. Therm. Eng.* **2022**, *209*, 118298.
18. Kharangate, C.R.; Mudawar, I. Review of computational studies on boiling and condensation. *Int. J. Heat Mass Transf.* **2017**, *108*, 1164–1196. [CrossRef]
19. Son, G.; Dhir, V.K.; Ramanujapu, N. Dynamics and Heat Transfer Associated With a Single Bubble during Nucleate Boiling on a Horizontal Surface. *ASME J. Heat Transf.* **1999**, *121*, 623–632.
20. Mukherjee, A.; Dhir, V.K. Study of lateral merger of vapor bubbles during nucleate pool boiling. *J. Heat Transf.* **2004**, *126*, 1023–1039.
21. Mukherjee, A.; Kandlikar, S.G. Numerical study of single bubbles with dynamic contact angle during nucleate pool boiling. *Int. J. Heat Mass Transf.* **2007**, *50*, 127.
22. Duan, X.; Phillips, B.; McKrell, T.; Buongiorno, J. Synchronized High-Speed Video, Infrared Thermometry, and Particle Image Velocimetry Data for Validation of Interface-Tracking Simulations of Nucleate Boiling Phenomena. *Exp. Heat Transf.* **2013**, *26*, 169–197.
23. Kunkelmann, C.; Stephan, P. CFD simulation of boiling flows using the volume of fluid method within OpenFOAM. *Numer. Heat Transf. A Appl.* **2009**, *56*, 631–646. [CrossRef]
24. Jiaqiang, E.; Zhang, Z.; Tu, Z.; Zuo, W.; Hu, W.; Han, D.; Jin, Y. Effect analysis on flow and boiling heat transfer performance of cooling water-jacket of bearing in the gasoline engine turbocharger. *Appl. Therm. Eng.* **2018**, *130*, 754–766.
25. Sun, D.L.; Tao, W.Q. A coupled volume-of-fluid and level set (VOSET) method for computing incompressible two-phase flows. *Int. J. Heat Mass Transf.* **2010**, *53*, 645–655. [CrossRef]
26. Ling, K.; Li, Z.Y.; Tao, W.Q. A direct numerical simulation for nucleate boiling by the VOSET method. *Numer. Heat Transf. A Appl.* **2014**, *65*, 949–971. [CrossRef]
27. She, X.; Shedd, T.A.; Lindeman, B.; Yin, Y.; Zhang, X. Bubble formation on solid surface with a cavity based on molecular dynamics simulation. *Int. J. Heat Mass Transf.* **2016**, *95*, 278–287.
28. Chen, Y.; Zou, Y.; Sun, D.; Wang, Y.; Yu, B. Molecular dynamics simulation of bubble nucleation on nanostructure surface. *Int. J. Heat Mass Transf.* **2018**, *118*, 1143–1151.
29. Zhang, L.; Xu, J.; Liu, G.; Lei, J. Nucleate boiling on nanostructured surfaces using molecular dynamics simulations. *Int. J. Therm. Sci.* **2020**, *152*, 106325.
30. Succi, S. *The Lattice Boltzmann Equation for Fluid Dynamics and Beyond*; Oxford University Press: Oxford, MI, USA, 2001; p. 304.
31. Chen, S.; Doolen, G.D. Lattice Boltzmann method for fluid flows. *Ann. Rev. Fluid Mech.* **1998**, *30*, 329–364.
32. Shan, X.; Chen, H. Lattice Boltzmann model for simulating flows with multiple phases and components. *Phys. Rev. E* **1993**, *47*, 1815–1820. [CrossRef] [PubMed]
33. Guo, Z.; Zheng, C.; Shi, B. Discrete lattice effects on the forcing term in the lattice Boltzmann method. *Phys. Rev. E* **2002**, *65*, 46308. [CrossRef] [PubMed]
34. He, X.; Shan, X.; Doolen, G. Discrete Boltzmann equation model for nonideal gases. *Phys. Rev. E Rapid Comm.* **1998**, *57*, 13. [CrossRef]
35. Kupershtokh, A.L. Incorporating a body force term into the lattice Boltzmann equation. *Vestnik NGU Quart. J. Novosibirsk State Univ. Ser. Math. Mech. Inform.* **2004**, *4*, 75–96.
36. Chen, S.; Martínez, D. On boundary conditions in lattice Boltzmann methods. *Phys. Fluid.* **1996**, *8*, 2527–2536. [CrossRef]

37. Li, Q.; Luo, K.H.; Kang, Q.J.; Chen, Q. Contact angles in the pseudopotential lattice Boltzmann modeling of wetting. *Phys. Rev. E* **2014**, *90*, 053301.
38. Gong, S.; Cheng, P. Lattice Boltzmann simulation of periodic bubble nucleation, growth and departure from a heated surface in pool boiling. *IJHMT* **2013**, *64*, 122–132. [CrossRef]
39. Fedoseev, A.V.; Surtaev, A.S.; Moiseev, M.I.; Ostapchenko, A.E. Lattice Boltzmann simulation of bubble evolution at boiling on surfaces with different wettability. *J. Phys. Conf. Ser.* **2020**, *1677*, 012085.
40. Gong, S.; Cheng, P. Lattice Boltzmann simulations for surface wettability effects in saturated pool boiling heat transfer. *Int. J. Heat Mass Transf.* **2015**, *85*, 635–646. [CrossRef]
41. Moiseev, M.I.; Fedoseev, A.; Shugaev, M.V.; Surtaev, A.S. Hybrid thermal lattice Boltzmann model for boiling heat transfer on surfaces with different wettability. *Int. Phenom. Heat Transf.* **2020**, *8*, 81. [CrossRef]
42. Li, Q.; Kang, Q.J.; Francois, M.M.; He, Y.L.; Luo, K.H. Lattice Boltzmann modeling of boiling heat transfer: The boiling curve and the effects of wettability. *Int. J. Heat Mass Transf.* **2015**, *85*, 787–796. [CrossRef]
43. Fang, W.Z.; Chen, L.; Kang, Q.J.; Tao, W.Q. Lattice Boltzmann modeling of pool boiling with large liquid-gas density ratio. *Int. J. Therm. Sci.* **2017**, *114*, 172–183. [CrossRef]
44. Zhang, C.; Cheng, P. Mesoscale simulations of boiling curves and boiling hysteresis under constant wall temperature and constant heat flux conditions. *Int. J. Heat Mass Transf.* **2017**, *110*, 319. [CrossRef]
45. Gong, S.; Cheng, P. Numerical simulation of pool boiling heat transfer on smooth surfaces with mixed wettability by lattice Boltzmann method. *Int. J. Heat Mass Transf.* **2015**, *80*, 206–216. [CrossRef]
46. Li, Q.; Yu, Y.; Zhou, P.; Yan, H.J. Enhancement of Boiling Heat Transfer Using Hydrophilic-Hydrophobic Mixed Surfaces: A Lattice Boltzmann Study. *Appl. Therm. Eng.* **2018**, *132*, 490–499. [CrossRef]
47. Peng, Y.; Laura, S. Equations of state in a lattice Boltzmann model. *Phys. Fluids* **2006**, *18*, 042101.
48. Nam, Y.; Wu, J.; Warrier, G.; Ju, Y.S. Experimental and numerical study of single bubble dynamics on a hydrophobic surface. *J. Heat Transf.* **2009**, *131*, 121004. [CrossRef]
49. Kim, S.H.; Lee, G.C.; Kang, J.Y.; Moriyama, K.; Park, H.S.; Kim, M.H. The role of surface energy in heterogeneous bubble growth on ideal surface. *Int. J. Heat Mass Transf.* **2017**, *108*, 1901–1909. [CrossRef]
50. Jo, H.; Ahn, H.S.; Kang, S.; Kim, M.H. A study of nucleate boiling heat transfer on hydrophilic, hydrophobic and heterogeneous wetting surfaces. *Int. J. Heat Mass Transf.* **2011**, *54*, 5643. [CrossRef]
51. Choi, C.H.; David, M.; Gao, Z.; Chang, A.; Allen, M.; Wang, H.; Chang, C.H. Large-scale Generation of Patterned Bubble Arrays on Printed Bi-functional Boiling Surfaces. *Sci. Rep.* **2016**, *6*, 23760. [CrossRef]
52. Nimkar, N.D.; Bhavnani, S.H.; Jaeger, R.C. Effect of nucleation site spacing on the pool boiling characteristics of a structured surface. *Int. J. Heat Mass Transf.* **2006**, *49*, 2829. [CrossRef]
53. Lim, D.Y.; Bang, I.C. Controlled bubble departure diameter on biphilic surfaces for enhanced pool boiling heat transfer performance. *Int. J. Heat Mass Transf.* **2020**, *150*, 119360. [CrossRef]

energies

MDPI

Article

The Influence of Pressure on Local Heat Transfer Rate under the Vapor Bubbles during Pool Boiling

Vladimir Serdyukov *, Ivan Malakhov and Anton Surtaev

Kutateladze Institute of Thermophysics, 630090 Novosibirsk, Russia
* Correspondence: vsserd@gmail.com

Abstract: This paper presents the results of an experimental study on the evolution of a nonstationary temperature field during ethanol pool boiling in a pressure range of 12–101.2 kPa. Experimental data were obtained using infrared thermography with high temporal and spatial resolutions, which made it possible to reconstruct the distribution of the heat flux density and to study the influence of pressure reduction on the local heat transfer rate in the vicinity of the triple contact line under vapor bubbles for the first time. It is shown that, for all studied pressures, a significant heat flux density is removed from the heating surface due to microlayer evaporation, which exceeds the input heat power by a factor of 3.3–27.7, depending on the pressure. Meanwhile, the heat transfer rate in the area of the microlayer evaporation significantly decreases with the pressure reduction. In particular, the local heat flux density averaged over the microlayer area decreases by four times as the pressure decreases from 101.3 kPa to 12 kPa. Estimates of the microlayer profile based on the heat conduction equation were made, which showed the significant increase in the microlayer thickness with the pressure reduction.

Keywords: boiling; heat transfer; microlayer; infrared thermography; subatmospheric pressures

Citation: Serdyukov, V.; Malakhov, I.; Surtaev, A. The Influence of Pressure on Local Heat Transfer Rate under the Vapor Bubbles during Pool Boiling. *Energies* **2023**, *16*, 3918. https://doi.org/10.3390/en16093918

Academic Editor: Moghtada Mobedi

Received: 14 April 2023
Revised: 30 April 2023
Accepted: 3 May 2023
Published: 5 May 2023

1. Introduction

Boiling is one of the most efficient heat transfer regimes, widely used in various technologies. To date, quite a large number of experimental and theoretical studies devoted to various aspects of boiling have been presented. Despite this, today there is no complete theory of heat transfer during nucleate boiling [1–3]. In particular, to calculate the heat transfer coefficients (HTC) during the boiling of various liquids under various conditions, dozens of semi-empirical correlations are given in the literature, which usually describe only a narrow change range of regime parameters. This is due to the fact that boiling is a multiscale nonstationary process, and for its correct description it is necessary to take into account the effects that take place on various scales, down to micro- and even nanoscales (Figure 1).

One of the phenomena that occur on the microscale during boiling is the formation and further evaporation of a liquid layer under a vapor bubble, which is called a microlayer due to its small thickness [4–6]. As can be seen from Figure 1, this region is limited on one side by the wall–liquid–vapor contact line and the surrounding liquid (the so-called macrolayer) on the other. To date, it has generally been accepted that the process of formation and evaporation of a microlayer is an essential and important part of a vapor bubble dynamics [7–9]. Moreover, it was shown in a number of papers [10–13] that the evaporation of a microlayer is one of the key mechanisms explaining the high intensity of heat transfer during nucleate boiling. Finally, understanding the nature and influence of various parameters on the microlayer evaporation rate is important to develop and fabricate micro- and nanostructured heating surfaces to achieve maximum heat transfer rates and critical heat fluxes during boiling [14–16].

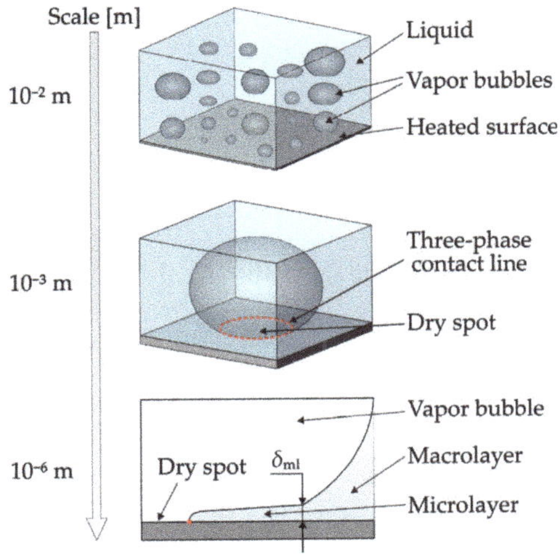

Figure 1. Multiscale nature of the nucleate boiling.

However, despite the rather large number of papers devoted to both the experimental study on microlayer evolution and structure [5–12] and the numerical simulation of its evaporation [17–20], a number of questions remain open. One of such issues is the influence of system parameters (pressure, subcooling degree, gravity level, etc.), as well as the properties of a heating surface (wettability, capillary wicking, structure, etc.) on the evolution and structure of a microlayer.

Pressure is one of the most important parameters in a boiling system, the change in which has a complex effect on the nucleation, dynamics of vapor bubbles, heat transfer rate and the development of boiling crisis phenomena. At the same time, it is known that the pressure reduction below atmospheric level leads not only to a quantitative, but also to a qualitative change in the boiling character [21–25]. In particular, a decrease in pressure is accompanied by a decrease in the intensity of heat transfer and a decrease in the critical heat flux, the nucleation of bubbles becomes more unstable and takes place at larger surface superheating. As a result, a noticeable change in the local characteristics of boiling (bubble growth rates, departure diameters, nucleation frequency, etc.) is observed. However, as the literature analysis shows, in contrast to boiling at atmospheric pressure, today there is practically no information about the features of local heat transfer and the dynamics of the contact line under vapor bubbles during boiling at subatmospheric pressures. Among other things, this is due to the fact that in most experimental studies the analysis of the temperature of a heating surface during boiling was carried out using thermocouples or temperature sensors. This method of temperature measurement can only be used to analyze the surface temperature averaged over the heater area.

Today, to measure the temperature field distribution of various objects, including boiling systems, the method of high-speed infrared thermography is widely used, which is devoid of the above-mentioned disadvantages. As an analysis of the papers of various authors (e.g., [10,23,26–34]) shows, this technique makes it possible to obtain fundamentally new information about the local and integral multiscale characteristics of boiling, as well as about the development of crisis phenomena under various conditions. In particular, infrared thermography allows the features of heat transfer in the vicinity of the triple contact line to be studied, including the evaporation of the microlayer during boiling [10,11].

In the present paper, an experimental study on the evolution of a nonstationary temperature field during the pool boiling of ethanol at pressures 12–101.2 kPa was performed. Experimental data were obtained using infrared thermography with high temporal and spatial resolutions, which made it possible to reconstruct the distribution of the heat flux density and to study the influence of pressure reduction on the local heat transfer rate in the vicinity of the triple contact line under the vapor bubbles.

2. Materials and Methods

2.1. Experimental Setup

The experiments were performed using the setup, a schematic view of which is shown in Figure 2. The setup consists of two sealed parts. To maintain a constant temperature of the working fluid, the inner volume was placed in thermostatic chamber filled with deionized water. The water temperature in the thermostatic volume was set at the required level using the Danfoss EKC-102 electronic temperature controller and two pre-heaters, each with a power of 1.2 kW. The temperature of the liquid in the inner volume (T_1) and the outer chamber (T_2) was monitored using Honeywell HEL 700 and NTC thermistors, respectively. In order to avoid the system pressure increase in the working area during boiling experiments, the internal volume was equipped with a water-cooled vapor condenser.

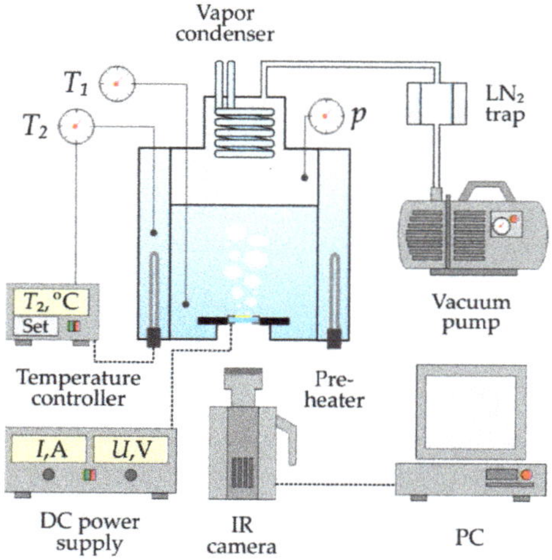

Figure 2. The scheme of the experimental setup.

To perform boiling experiments under subatmospheric pressure conditions, the setup was evacuated according to DIN 28400-1-1990 norm. The required pressure level in the working volume was set using a rotary vane vacuum pump EVP 2XZ-1C, equipped with a liquid nitrogen trap, and controlled using a digital piezoresistive vacuum gauge Thyracont VD81. The value of the reduced pressure p_s in the experiments was calculated based on the readings of the vacuum gauge (p), taking into account the influence of the hydrostatic pressure of the working fluid column with height h:

$$p_s = p + \rho_l g h \quad (1)$$

As the working fluid, 95% ethanol was used under the saturation conditions at a given pressure p_s. During experiments on boiling at low pressures, it is extremely important to correctly calculate the system conditions. In the present study, for such conditions (p_s, T_{sat}), the ones near a heating surface were considered. To do this, in each experiment the value of the reduced pressure was calculated according to (1) and the corresponding saturation temperature of the working fluid T_{sat} was taken from the tabular data. The following reduced pressures were studied in the paper: 12 kPa (T_{sat} = 32.4 °C), 21.8 kPa (T_{sat} = 44.3 °C), 35 kPa (T_{sat} = 53.6 °C), 57.6 kPa (T_{sat} = 64.7 °C), 80 kPa (T_{sat} = 72.4 °C) and 101.3 kPa (T_{sat} = 78.6 °C).

2.2. Heating Surface

The heating surface used in the study (Figure 3) was a conductive film of indium-tin oxide (ITO), 1 μm thick, deposited by ion sputtering onto a sapphire substrate 60 mm in diameter and 3 mm thick. The area of the heat release was 30 × 28 mm^2. Samples were resistively heated by a DC power supply Elektro Automatik PS 8080-60 DT via thin (2 μm) chromium-nickel electrodes vacuum deposited onto the ITO film and brought to the reverse side of the sapphire substrate.

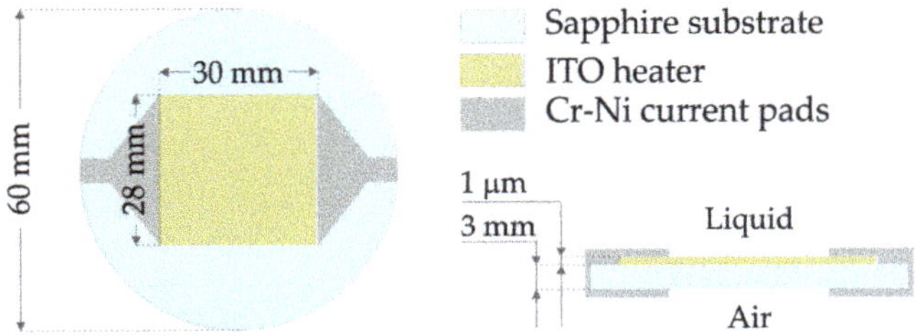

Figure 3. The scheme of the heating surface used in the study.

An important advantage of using indium–tin oxide as a heater material in experiments to study boiling characteristics is its transparency in the visible wavelength range (380–750 nm) and opacity in the mid-IR range (3–5 μm). In turn, the integral transmission capacity of a sapphire in the 0.3–5 μm wavelength range exceeds 80%. The combination of these properties makes it possible to measure the non-stationary temperature field on the ITO film using an infrared camera with corresponding spectral range and to visually record the vapor bubble dynamics on the heating surface using a video camera [23,26–31].

In the present study, the transient temperature field of the ITO heater was recorded using a FLIR X6530sc high-speed infrared camera with a spectral range of 1.5–5.1 μm, 160 × 128 resolution and a frame rate of 1500 fps. The spatial resolution of the IR recording in the experiments was 250 μm per pixel. Prior to the boiling experiments, the IR camera was calibrated using a resistance temperature detector located near the film heater [23].

Prior to the experiments, an analysis of the physicochemical properties of the heat-releasing surface was performed. According to the profilometry performed using the Bruker Contour GT-K1 optical profilometer, the ITO film roughness was R_a = 3 nm, which classifies it as an "ultra" smooth heating surface. Using the KRUSS DSA 100 setup, an analysis of the wetting properties of the ITO film was performed for the working fluid to determine the value of the static contact angle (θ). According to these measurements, the value of θ was about 10°. To additionally check the working fluid for the absence of harmful insoluble impurities, the wetting properties of the heating surface were also investigated after boiling experiments. An analysis of these results showed that the measured values of the contact wetting angles before and after the experiments are consistent with each other within the error (±2°), which indicates sufficient purity of the used liquid and the absence of harmful impurities in it.

2.3. Local Heat Flux Calculation

To analyze the local density of the heat flux coming from the heater to the liquid, a numerical calculation was performed using the experimental data of infrared thermography on the evolution of the ITO film temperature field. Below, the problem of heat conduction to recover local heat fluxes will be described. In the study, a numerical solution of the non-stationary heat conduction equation was performed, which has the following form:

$$\frac{\partial T}{\partial t} = \lambda_{\text{sap}}\left(\frac{\partial^2 T}{\partial x^2} + \frac{\partial^2 T}{\partial y^2} + \frac{\partial^2 T}{\partial z^2}\right). \quad (2)$$

The three-dimensional internal volume of the sapphire substrate (Figure 4a) was used as the computational domain, bounded from above and below by the outer boundaries of the sapphire, and from the sides by the boundaries of the heat release region. The boundary conditions were given as follows:

$$T(x, y)_{z=z_{\max}} = T_{IR}(x, y), \tag{3}$$

$$\frac{\partial T}{\partial z}_{z=0} = 0,\ \frac{\partial T}{\partial x}_{x=0, x=x_{\max}} = 0,\ \frac{\partial T}{\partial y}_{y=0. y=y_{\max}} = 0. \tag{4}$$

These conditions correspond to the absence of heat losses to the sides, which, as the analysis of the experimental results showed, is quite correct. The temperature field of the wall–liquid contact surface changed with time, as happened in the experiment.

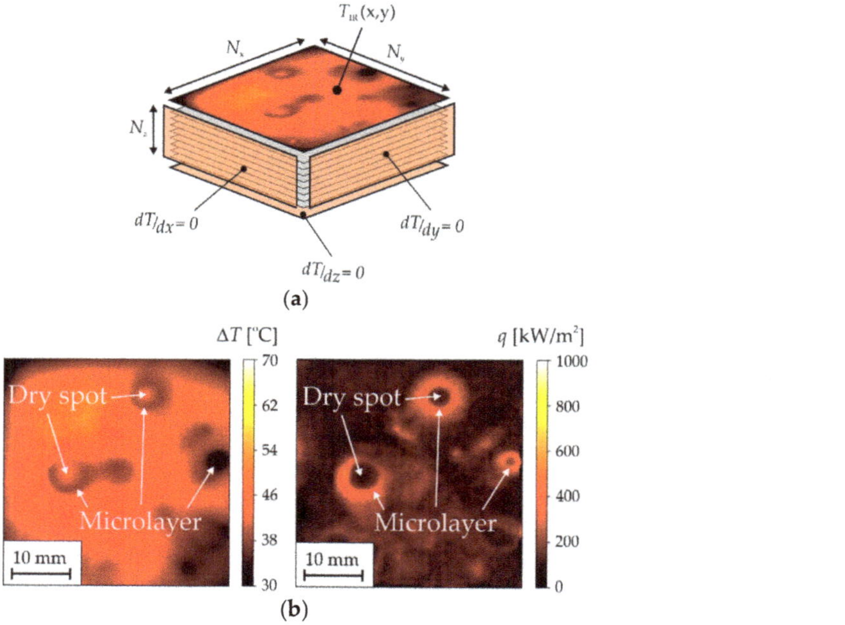

Figure 4. (**a**) The calculation domain including boundary conditions to reconstruct heat flux density distribution maps during boiling on the ITO heater based on the infrared thermography data; (**b**) an example of the reconstruction of the heat flux distribution map (p_s = 21.8 kPa, q_{input} = 150 kW/m^2).

The calculations were performed using an explicit finite difference scheme. The splitting of the calculated area along the X and Y axes corresponded to the splitting of the IR frame to the pixels. The splitting along the Z axis and in time were chosen based on the analysis of the decrease in the calculation error with an increase in the amount of splitting. For calculations at intermediate moments between recorded IR frames, the temperature field was recovered using linear interpolation. The temperature field measured with IR camera was set as the initial temperature distribution in the calculated volume on all layers.

As a result of the numerical calculations, the non-stationary temperature distribution in the volume of the sapphire substrate was recovered. To calculate the density of the heat flux incoming to the liquid, the energy conservation condition was used, which takes into account the heat release in the ITO film q_{input}:

$$q_l = q_{\text{input}} - q_{\text{sap}}, \tag{5}$$

where q_{sap} is the heat flux density going into the sapphire substrate, determined on the basis of the calculated temperature distribution in the volume of sapphire:

$$q_{sap} = \lambda \frac{\partial T}{\partial z} \quad (6)$$

As an example, Figure 4b shows an IR recording frame and the corresponding calculated distribution of the heat flux density during the boiling of ethanol at p_s = 21.8 kPa (q_{input} = 150 kW/m^2). As can be seen, the obtained experimental data and the heat fluxes map calculated on their basis make it possible to identify and analyze various heat transfer areas in the vicinity of the contact line (Figure 1). In particular, the area in the center of the vapor bubble with a higher temperature and with a lower heat flux density consequently is a dry spot bounded by a triple contact line. In turn, the region along the periphery of the bubble with the minimum temperature and the maximum heat flux density corresponds to the region of a microlayer evaporation.

2.4. Measurements Uncertainties

The uncertainty of the pressure measurement consists of the sum of the pressure sensor accuracy and the accuracy of the liquid column height h measurement. The total pressure measurement error was about 1% in the experiments. In turn, the uncertainty of the input heat flux measurement includes inaccuracies associated with the current, voltage gauging, and heat losses. According to the apparatus data sheets, the total error in current and voltage measurements is no more than 1%. In turn, the 2D numerical calculation in Comsol Multiphysics was performed to analyze the lateral heat losses for the steady-state conditions. The results showed that the heat losses contribute about 5%, since the sapphire substrate has a relatively low thermal conductivity and the heat transfer coefficients on the ITO-liquid surface are rather high. Therefore, the total uncertainty of the heat flux density measurement in the experiments was about 6%. The major contributions to the uncertainty of the surface temperature measurements using IR camera are an uncertainty associated with the calibration procedure and camera sensitivity. According to the analysis, the total uncertainty of the surface temperature measurement was no more than 1.5 K.

To verify the correctness of the performed data gathering and curation, a comparison of the obtained results on the heat transfer rate during boiling at atmospheric pressure with the models of Rohsenow, 1952 [35] and Yagov, 1988 [36] was made (Figure 5). As the literature analysis shows, these models are some of the most used to describe boiling data for various liquids and heating surfaces.

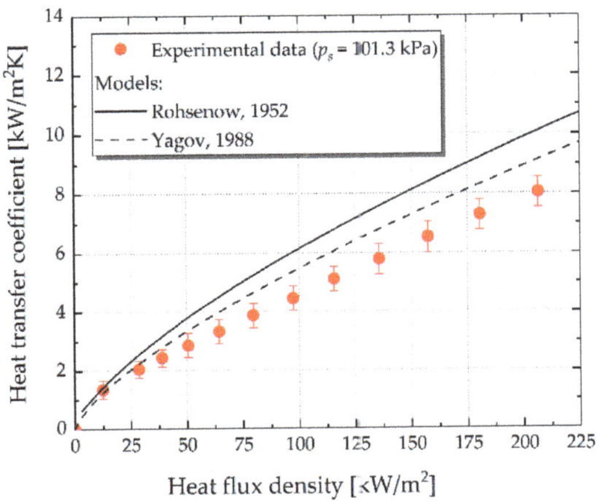

Figure 5. Comparison of the obtained experimental data on the heat transfer rate during ethanol pool boiling under atmospheric pressure with the models of Rohsenow (1952) [35] and Yagov (1988) [36].

As can be seen from Figure 5, the experimental data show the same trend as the used dependencies, which confirms that the data were collected correctly. However, it is also seen that the experiments show lower values of HTC compared to the calculations. Apparently, this is due to a rather low roughness of the heating surface used in the study. In particular, it is well known [37] that smooth and ultrasmooth surfaces are characterized by much lower heat transfer rates during boiling compared to the so-called technical surfaces, for which the semi-empirical models shown in Figure 5 were developed.

3. Results

3.1. Heat Transfer Rate

First of all, the effect of the pressure reduction on the heat transfer rate during ethanol boiling was studied. To do this, the values of the integral temperature of the heating surface (averaged over the area and recording time (10 s)) were obtained based on the IR recording data. As the result, the corresponding boiling curves $q(\Delta T)$ were plotted for various pressures (Figure 6a). As can be seen from the figure, the pressure reduction in a given range is accompanied by a significant increase in the onset of nucleate boiling (ONB). In particular, if for atmospheric pressure conditions the onset of nucleate boiling corresponds to surface superheating ΔT = 14 K, then for the lowest studied pressure (12 kPa) this value is 37 K.

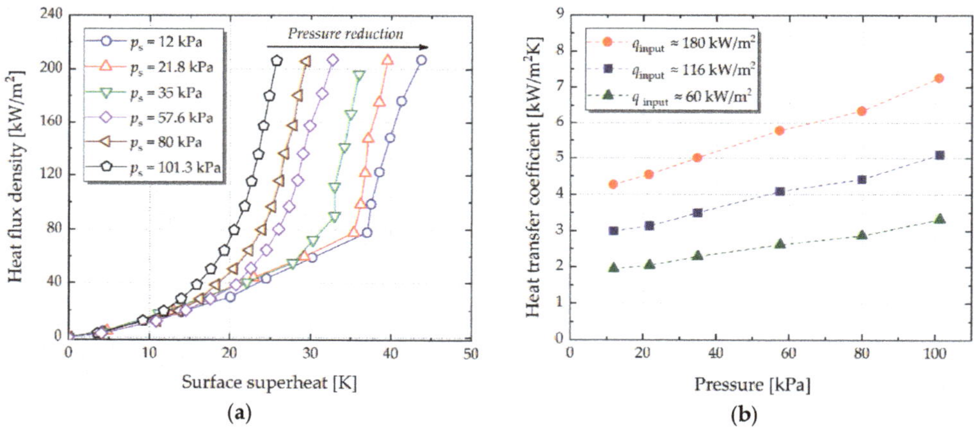

Figure 6. (**a**) Ethanol boiling curves at various pressures p_s; (**b**) the pressure influence on the heat transfer coefficient during ethanol boiling at various heat flux densities.

Using the obtained data, the heat transfer coefficients HTC were also determined and dependencies of HTC(p_s) were constructed for different heat flux densities q_{input} (Figure 6b). An analysis of the obtained curves shows that for all presented heat flux densities, the dependence of HTC(p_s) has a linear form. The pressure reduction from 101.3 kPa to 12 kPa leads to the heat transfer deteriorating by approximately 40%. The literature analysis shows that the results on the influence of pressure reduction on heat transfer rate during boiling obtained in the present study are consistent with the data of other authors obtained using different liquids, including ethanol [38].

As noted in the introduction, such a decrease in the heat transfer rate during boiling with the pressure reduction is directly related to a significant change in the nucleation and vapor bubbles dynamics. This is clearly demonstrated by the analysis of heat flux distribution maps obtained according to the algorithm described in Section 2.3 (Figure 7).

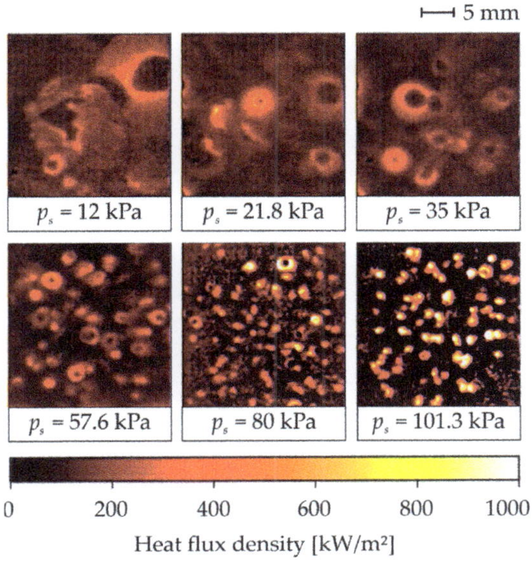

Figure 7. Heat flux density distribution maps during boiling of ethanol at various pressures.

In particular, Figure 7 demonstrates that a noticeable decrease in the number of nucleation sites happens with the pressure reduction, while the departure diameters of the vapor bubbles and their lifetime on the heating surface increase. In addition, it is clearly seen that in the area of microlayer evaporation, the local heat flux density decreases with pressure reduction. At the same time, it is also seen that a change in pressure affects the heat transfer rate in the regions free of vapor bubbles, i.e., the convective component of the heat transfer according to the mechanistic approach [39,40]. Thus, for atmospheric pressure, the average heat flux density removed by the convection does not exceed 50 kW/m^2, while for pressures below 35 kPa this value is about 150 kW/m^2, which is comparable to the input heat power. In the future, the authors plan to perform a more detailed analysis of the effect of pressure reduction on the contribution of various heat transfer mechanisms (e.g., convection, microlayer evaporation, and quenching) during boiling, based on the machine learning algorithms [41] of infrared thermography dataset analysis. Further, a more detailed analysis of the features of local heat transfer under single nucleation sites during boiling at various pressures will be presented.

3.2. Local Heat Transfer under the Vapor Bubbles

Figure 8 demonstrates the heat flux density distribution under a vapor bubble at different moments for various pressures and the same input heat flux density (q_{input} = 150 kW/m^2). It can be seen that for all pressures, the maximum local heat flux is observed in the region of microlayer evaporation at the initial stage of vapor bubble growth (t = 0.67–1.3 ms depending on pressure). Additionally, the appearance of a region with reduced heat transfer under the vapor bubble's center is clearly visible for all bubbles, which corresponds to the formation and growth of a dry spot on a heating wall. As a result, the local heat flux spent to evaporate the microlayer decreases as the bubble grows and becomes comparable to the input heat power at the final stage of the bubble's lifecycle. In addition, the presented figures reveal a significant effect of a pressure reduction on the complete depletion time of the microlayer, which is associated with a significant increase in the lifetime of a bubble with a decrease in pressure. In particular, for atmospheric pressure, the microlayer fully evaporated in 6 ms. In turn, as the pressure decreased to p_s = 12 kPa, this value increased to more than 26 ms.

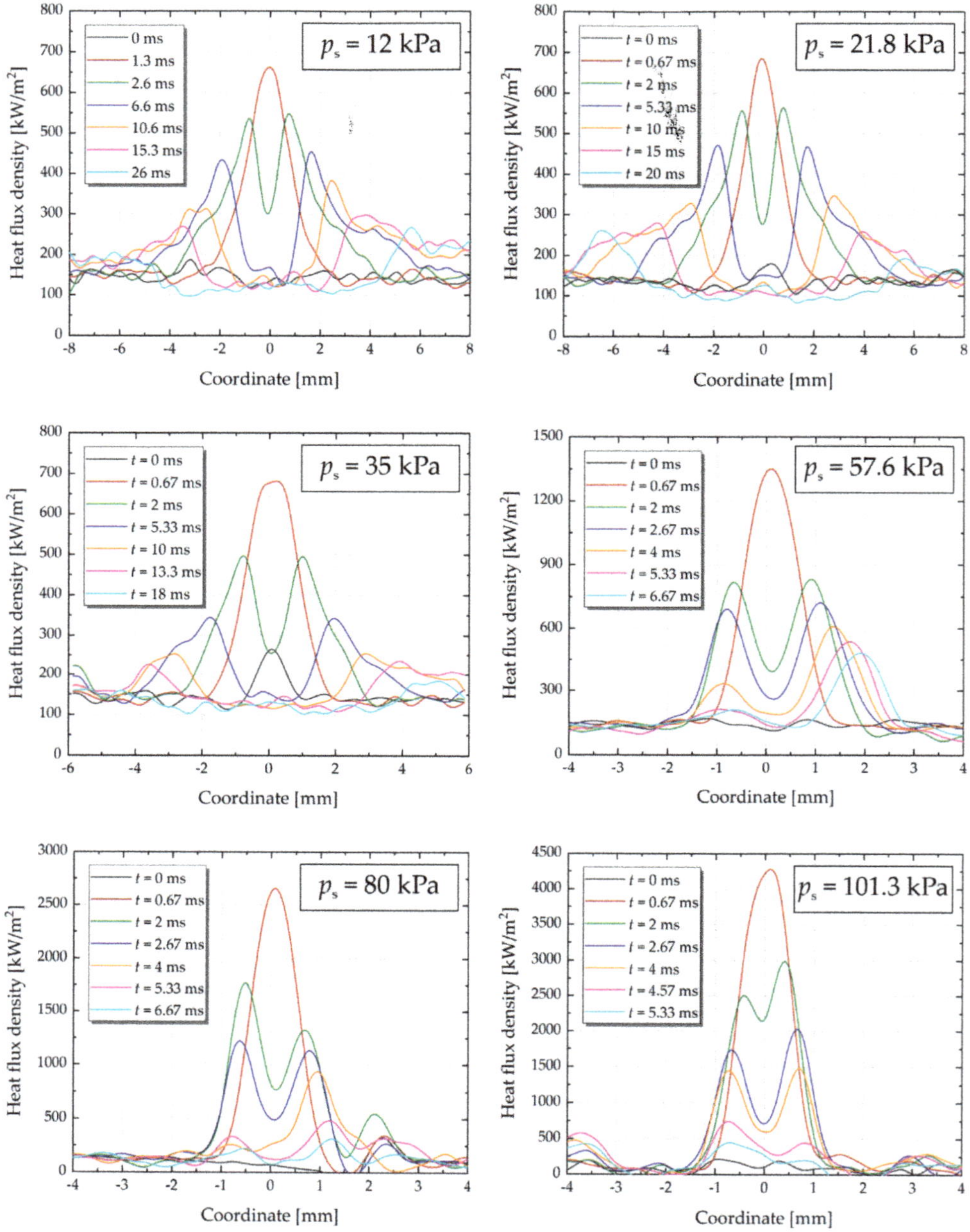

Figure 8. Evolution of the heat flux density distribution under vapor bubbles during ethanol boiling at various pressures p_s (q_{input} = 150 kW/m^2).

In the further analysis of the microlayer region, the following criteria were used to determine its boundaries: the inner microlayer boundary corresponds to the maximum intensity of local heat transfer. In turn, the outer boundary of the microlayer was determined from the condition that the local heat flux does not change with distance from the bubble center.

In addition, the presented figures show that the region of microlayer evaporation noticeably increases in size with pressure reduction. In particular, if its maximum transverse size during boiling at atmospheric pressure is about 2.5 mm, then at 12 kPa this value is more than 10 mm. On the one hand (also taking into account the longer lifetime of the microlayer), this suggests that during boiling at low pressures, the evaporation of the microlayer should make a greater contribution to the overall heat transfer compared to boiling at atmospheric conditions. On the other hand, the results obtained indicate a noticeable decrease in the density of the heat flux removed from the heating surface due to the microlayer evaporation with pressure reduction. As noted earlier, the authors plan to use the obtained data for a more accurate analysis of the effect of pressure reduction on the contribution of various mechanisms, including microlayer evaporation, to the heat transfer during boiling.

An analysis of the results shows that with pressure reduction, the maximum heat flux density spent to microlayer evaporation decreases significantly (Figure 9a). Thus, the ratio of this value ($q_{\mathrm{ml\,max}}$) to the input heat flux density decreases from 27.7 times ($q_{\mathrm{ml\,max}} \approx 4.36 \times 10^3$ kW/m^2) at boiling under atmospheric pressure to 3.3 times ($q_{\mathrm{ml\,max}} \approx 484$ kW/m^2) at $p_s = 12$ kPa.

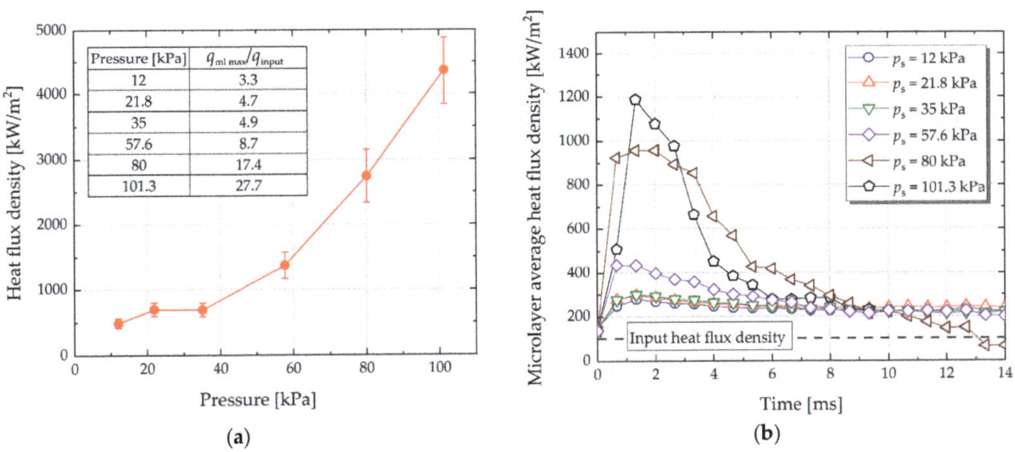

Figure 9. (**a**) Dependence of the maximum heat flux density in the microlayer evaporation area on pressure; (**b**) evolution of the local heat flux density averaged over the microlayer area during ethanol boiling at various pressures ($q_{\mathrm{input}} = 150$ kW/m^2).

Additionally, two trends are observed in Figure 9a. If the pressure reduction from atmospheric to 35 kPa demonstrates close to linear dependence $q_{\mathrm{ml\,max}}(p_s)$, then a further decrease in pressure to 12 kPa no longer leads to such a significant drop in the heat transfer intensity in the microlayer evaporation region. It is interesting to note that in the previous studies of the authors [23], a change in the dynamics of the contact line under vapor bubbles was also experimentally found during water boiling at a similar pressure. In particular, the authors showed that if pressure reduction to 42 kPa leads to a decrease in the growth rate of the dry spot (which is directly related to the evaporation of the microlayer), then a further pressure reduction, on the contrary, is characterized by an increasing growth rate of the dry spot. This indicates that, apparently, at these pressure levels, there is a change in the determining mechanisms of the intensity of microlayer evaporation and the dynamics of the triple contact line. Nevertheless, an accurate answer to this question requires a comprehensive analysis of both the dynamics of the contact line and the structure of the microlayer during the boiling of various liquids at various pressures.

Using the obtained experimental data, the evolution of the local heat flux averaged over the area of microlayer evaporation was also studied for various pressures (Figure 9b).

It can be seen that for all studied pressures, the main contribution to heat transfer is made by the evaporation of the microlayer at the initial stage of vapor bubble growth. This figure again indicates that the pressure reduction leads to a noticeable decrease in the heat flux density spent to evaporate the microlayer. In particular, the heat flux density averaged over the microlayer area decreases by four times with a pressure reduction from atmospheric level to 12 kPa.

3.3. Microlayer Thickness Estimation

Using the obtained data on the heat flux densities spent to evaporate the microlayer, it becomes possible to reconstruct its profile $\delta_{\text{ml}}(R_b)$, where δ_{ml} is the thickness of the microlayer at a distance R_b from the center of a bubble. A fairly simple estimate can be made using the following heat conduction equation:

$$q_{\text{ml}} = \frac{k_l}{\delta_{\text{ml}}}(T_w - T_i) \tag{7}$$

where T_w is the temperature of a heating surface and T_i is the temperature at the vapor–liquid interface. In this case, the temperature T_i was taken to be the saturation temperature for a given pressure. Figure 10 shows the microlayer profiles at the initial moment of its formation for pressures p_s = 12, 57.6 and 101.3 kPa. Additionally, the values of the so-called initial thickness of the microlayer $\delta_{\text{ml}\,0}$ (Figure 1) are presented. The obtained profiles show that with a pressure reduction from atmospheric pressure to the lowest one, the microlayer becomes more than eight times thicker. Moreover, it can be seen that at p_s = 12 kPa, the minimum thickness of the microlayer in the vicinity of the triple contact line is noticeably larger and is about 10 μm. This may be due to a significant change in the shape of vapor bubbles during boiling at subatmospheric pressures and their strong deviation from the spherical shape characteristic of boiling at atmospheric pressure. In particular, a number of authors showed via high-speed video recordings [22,23,25] that the vapor bubbles formed during boiling in vacuum have a "mushroom" shape with a pronounced vapor stem, which connects the bubble body with a heating surface.

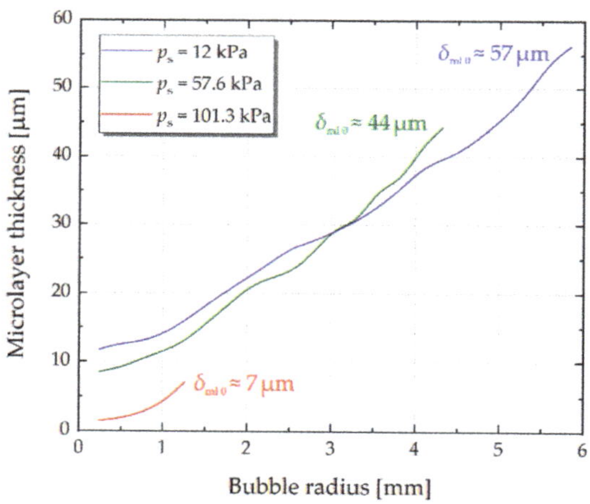

Figure 10. Microlayer profiles during ethanol boiling at various pressures.

It is important to note here that the performed analysis is only an estimate of the microlayer thickness during boiling at various pressures, which does not claim to be quantitatively accurate. In particular, a number of authors [42,43] showed that the heat flux spent to evaporate the microlayer depends significantly on the so-called evaporation resistance associated with the molecular dynamics of the evaporation process at the liquid–vapor interface. Ignoring it and calculating the microlayer thickness only using the heat conduction Equation (7) shows overestimated values of $\delta_{ml}(R_b)$. For an accurate description of the microlayer profile, it is necessary to perform direct experimental observations using interferometric methods [6,7,10,12], which make it possible to study the process of its formation and growth with high temporal and spatial resolutions and to study the $\delta_{ml}(R_b)$ dependence in detail. Here, the method of LED interferometry proposed recently in [44] should be especially noted, which is a simpler alternative to laser interferometry for studying the structure and evolution of a microlayer during boiling. Using the example of water flow boiling, the authors of [44] showed that this relatively easy-to-implement technique makes it possible to obtain distinct interference patterns during the formation and growth of bubbles, which makes it possible to analyze the microlayer thickness. The usage of LED interferometry simultaneously with the high-speed infrared thermography is promising for studying the effect of pressure on the microlayer structure during boiling. The authors of the present paper plan to perform such a study in the near future.

4. Conclusions

In the present paper, the effect of pressure reduction on the local heat transfer in the vicinity of the contact line during pool boiling under conditions was studied for the first time. Experimental data on the evolution of the non-stationary temperature field of the heating surface during the ethanol boiling in the pressure range of 12–101.3 kPa were obtained using high-speed infrared thermography. The developed numerical algorithm made it possible to reconstruct the distribution of the heat flux density over the heater for various pressures and to study the heat transfer in the area of microlayer evaporation. The analysis of the data showed the following:

- The dependence of heat transfer coefficients on pressure during ethanol pool boiling has a linear form. A decrease in pressure from 101.3 kPa to 12 kPa leads to a decrease in the intensity of heat transfer of about 40%. At the same time, the onset of nucleate boiling noticeably increases—from 14 to 37 K.
- The data obtained indicate the effect of pressure reduction on the contribution of various mechanisms to the integral heat transfer rate during boiling. The results can be further used to perform a more accurate analysis of this influence, including using machine learning algorithms.
- For all studied pressures, a significant heat flux density is removed from the heating surface due to microlayer evaporation ($q_{ml\,max}$ = 484 kW/m^2–4.36 $\times$ 10^3 kW/m^2), which exceeds the input heat power by a factor of 3.3–27.7, depending on pressure.
- The heat transfer rate in the area of the microlayer evaporation significantly decreases with reduction in pressure. Thus, local heat flux density averaged over the microlayer area decreases by four times as the pressure decreases from 101.3 kPa to 12 kPa.
- Estimates of the microlayer thickness based on the heat conduction equation were made. The results showed that the pressure reduction from atmospheric level to 12 kPa leads to the microlayer thickness increasing by more than eight times—from 7 µm to 57 µm.

Author Contributions: Conceptualization, V.S. and A.S.; methodology, V.S. and A.S.; software, I.M.; validation, V.S. and I.M.; formal analysis, V.S. and I.M.; investigation, I.M. and V.S.; resources, V.S. and A.S.; data curation, V.S. and I.M.; writing—original draft preparation, V.S.; writing—review and editing, V.S.; visualization, V.S. and I.M.; supervision, V.S.; project administration, V.S.; funding acquisition, V.S. All authors have read and agreed to the published version of the manuscript.

Funding: The work was supported by the Russian Science Foundation (Grant No. 22-79-00174). The experimental setup was designed and made within the framework of the state assignment of the IT SB RAS (№ 121031800216-1).

Data Availability Statement: Not applicable.

Conflicts of Interest: The authors declare no conflict of interest.

References

1. Yagov, V.V. Nucleate boiling heat transfer: Possibilities and limitations of theoretical analysis. *Heat Mass Transf.* **2009**, *45*, 881–892. [CrossRef]
2. Koizumi, Y.; Shoji, M.; Monde, M.; Takata, Y.; Nagai, N. *Boiling: Research and Advances*; Elsevier: Amsterdam, The Netherlands, 2017.
3. Liang, G.; Mudawar, I. Pool boiling critical heat flux (CHF)—Part 2: Assessment of models and correlations. *Int. J. Heat Mass Transf.* **2018**, *117*, 1368–1383. [CrossRef]
4. Moore, F.D.; Mesler, R.B. The measurement of rapid surface temperature fluctuations during nucleate boiling of water. *AIChE J.* **1961**, *7*, 620–624. [CrossRef]
5. Cooper, M.G.; Lloyd, A.J.P. The microlayer in nucleate pool boiling. *Int. J. Heat Mass Transf.* **1969**, *12*, 895–913. [CrossRef]
6. Voutsinos, C.M.; Judd, R.L. Laser Interferometric Investigation of the Microlayer Evaporation Phenomenon. *ASME J. Heat Transf.* **1975**, *97*, 88–92. [CrossRef]
7. Jawurek, H.H. Simultaneous determination of microlayer geometry and bubble growth in nucleate boiling. *Int. J. Heat Mass Transf.* **1969**, *12*, 843–848. [CrossRef]
8. Chen, Z.; Utaka, Y. On heat transfer and evaporation characteristics in the growth process of a bubble with microlayer structure during nucleate boiling. *Int. J. Heat Mass Transf.* **2015**, *81*, 750–759. [CrossRef]
9. Srivastava, A. On the identification and mapping of three distinct stages of single vapor bubble growth with the corresponding microlayer dynamics. *Int. J. Multiph. Flow* **2021**, *142*, 103722.
10. Jung, S.; Kim, H. An experimental study on heat transfer mechanisms in the microlayer using integrated total reflection, laser interferometry and infrared thermometry technique. *Heat Transf. Eng.* **2015**, *36*, 1002–1012. [CrossRef]
11. Serdyukov, V.S.; Surtaev, A.S.; Pavlenko, A.N.; Chernyavskiy, A.N. Study on local heat transfer in the vicinity of the contact line under vapor bubbles at pool boiling. *High Temp.* **2018**, *56*, 546–552. [CrossRef]
12. Utaka, Y.; Hu, K.; Chen, Z.; Morokuma, T. Measurement of contribution of microlayer evaporation applying the microlayer volume change during nucleate pool boiling for water and ethanol. *Int. J. Heat Mass Transf.* **2018**, *125*, 243–247. [CrossRef]
13. Narayan, L.S.; Srivastava, A. Non-contact experiments to quantify the microlayer evaporation heat transfer coefficient during isolated nucleate boiling regime. *Int. Commun. Heat Mass Transf.* **2021**, *122*, 105191. [CrossRef]
14. Zou, A.; Singh, D.P.; Maroo, S.C. Early evaporation of microlayer for boiling heat transfer enhancement. *Langmuir* **2016**, *32*, 10808–10814. [CrossRef]
15. Ding, W.; Zhang, J.; Sarker, D.; Hampel, U. The role of microlayer for bubble sliding in nucleate boiling: A new viewpoint for heat transfer enhancement via surface engineering. *Int. J. Heat Mass Transf.* **2020**, *149*, 119239. [CrossRef]
16. Bongarala, M.; Hu, H.; Weibel, J.A.; Garimella, S.V. Microlayer evaporation governs heat transfer enhancement during pool boiling from microstructured surfaces. *Appl. Phys. Lett.* **2022**, *120*, 221602. [CrossRef]
17. Urbano, A.; Tanguy, S.; Huber, G.; Colin, C. Direct numerical simulation of nucleate boiling in micro-layer regime. *Int. J. Heat Mass Transf.* **2018**, *123*, 1128–1137. [CrossRef]
18. Guion, A.; Afkhami, S.; Zaleski, S.; Buongiorno, J. Simulations of microlayer formation in nucleate boiling. *Int. J. Heat Mass Transf.* **2018**, *127*, 1271–1284. [CrossRef]
19. Bureš, L.; Sato, Y. Comprehensive simulations of boiling with a resolved microlayer: Validation and sensitivity study. *J. Fluid Mech.* **2022**, *933*, A54. [CrossRef]
20. Lakew, E.; Sarchami, A.; Giustini, G.; Kim, H.; Bellur, K. Thin film evaporation modeling of the liquid microlayer region in a dewetting water bubble. *Fluids* **2023**, *8*, 126. [CrossRef]
21. Kutateladze, S.S.; Mamontova, N.N. Critical heat fluxes in the pool boiling of liquids at reduced pressure. *J. Eng. Phys.* **1967**, *12*, 86–90. [CrossRef]
22. Van Stralen, S.J.D.; Zijl, W.; De Vries, D.A. The behaviour of vapour bubbles during growth at subatmospheric pressures. *Chem. Eng. Sci.* **1977**, *32*, 1189–1195. [CrossRef]
23. Surtaev, A.; Serdyukov, V.; Malakhov, I. Effect of subatmospheric pressures on heat transfer, vapor bubbles and dry spots evolution during water boiling. *Exp. Therm. Fluid Sci.* **2020**, *112*, 109974. [CrossRef]
24. Emir, T.; Ourabi, H.; Budakli, M.; Arik, M. Parametric effects on pool boiling heat transfer and critical heat flux: A critical review. *J. Electron. Packag.* **2022**, *144*, 040801. [CrossRef]
25. Mahmoud, M.M.; Karayiannis, T.G. Bubble growth on a smooth metallic surface at atmospheric and sub-atmospheric pressure. *Int. J. Heat Mass Transf.* **2023**, *209*, 124103. [CrossRef]
26. Gerardi, C.; Buongiorno, J.; Hu, L.W.; McKrell, T. Study of bubble growth in water pool boiling through synchronized, infrared thermometry and high-speed video. *Int. J. Heat Mass Transf.* **2010**, *53*, 4185–4192. [CrossRef]
27. Su, G.Y.; Wang, C.; Zhang, L.; Seong, J.H.; Kommajosyula, R.; Phillips, B.; Bucci, M. Investigation of flow boiling heat transfer and boiling crisis on a rough surface using infrared thermometry. *Int. J. Heat Mass Transf.* **2020**, *160*, 120134. [CrossRef]
28. Surtaev, A.; Serdyukov, V.; Malakhov, I.; Safarov, A. Nucleation and bubble evolution in subcooled liquid under pulse heating. *Int. J. Heat Mass Transf.* **2021**, *169*, 120911. [CrossRef]
29. Ronshin, F.V.; Dementiev, Y.A.; Chinnov, E.A. Investigation of dielectric liquid FC-72 boiling in a slit microchannel. *Thermophys. Aeromech.* **2022**, *29*, 975–980. [CrossRef]
30. Surtaev, A.; Koşar, A.; Serdyukov, V.; Malakhov, I. Boiling at subatmospheric pressures on hydrophobic surface: Bubble dynamics and heat transfer. *Int. J. Therm. Sci.* **2022**, *173*, 107423. [CrossRef]

31. Kangude, P.; Srivastava, A. Experiments to understand bubble base evaporation mechanisms and heat transfer on nano-coated surfaces of varying wettability under nucleate pool boiling regime. *Int. J. Multiph. Flow* **2022**, *152*, 104098. [CrossRef]
32. Može, M.; Hadžić, A.; Zupančič, M.; Golobič, I. Boiling heat transfer enhancement on titanium through nucleation-promoting morphology and tailored wettability. *Int. J. Heat Mass Transf.* **2022**, *195*, 123161. [CrossRef]
33. Surtaev, A.; Malakhov, I.; Serdyukov, V. Explosive vaporization of ethanol on microheater during pulse heating. *Heat Transf. Eng.* **2022**, *44*, 502–511. [CrossRef]
34. Sielaff, A.; Mangini, D.; Kabov, O.; Raza, M.Q.; Garivalis, A.I.; Zupančič, M.; Dehaeck, S.; Evgenidis, S.; Jacobs, C.; Van Hoof, D.; et al. The multiscale boiling investigation on-board the International Space Station: An overview. *Appl. Therm. Eng.* **2022**, *205*, 117932. [CrossRef]
35. Rohsenow, W.M. A method of correlating heat-transfer data for surface boiling of liquids. *Trans. Am. Soc. Mech. Eng.* **1952**, *74*, 969–975. [CrossRef]
36. Yagov, V.V. Heat transfer with developed nucleate boiling of liquids. *Therm. Eng.* **1988**, *35*, 65.
37. Jones, B.J.; McHale, J.P.; Garimella, S.V. The influence of surface roughness on nucleate pool boiling heat transfer. *J. Heat Transf.* **2009**, *131*, 121009. [CrossRef]
38. Kalani, A.; Kandlikar, S.G. Enhanced pool boiling with ethanol at subatmospheric pressures for electronics cooling. *J. Heat Transf.* **2013**, *135*, 111002. [CrossRef]
39. Kurul, N.; Podowski, M.Z. Multidimensional effects in forced convection subcooled boiling. In Proceedings of the International Heat Transfer Conference Digital Library, Jerusalem, Israel, 19–24 August 1990; Begel House Inc.: Danbury, CT, USA, 1990.
40. Benjamin, R.J.; Balakrishnan, A.R. Nucleate pool boiling heat transfer of pure liquids at low to moderate heat fluxes. *Int. J. Heat Mass Transf.* **1996**, *39*, 2495–2504. [CrossRef]
41. Malakhov, I.; Seredkin, A.; Chernyavskiy, A.; Serdyukov, V.; Mullyadzanov, R.; Surtaev, A. Deep learning segmentation to analyze bubble dynamics and heat transfer during boiling at various pressures. *Int. J. Multiph. Flow* **2023**, *162*, 104402. [CrossRef]
42. Giustini, G.; Jung, S.; Kim, H.; Walker, S.P. Evaporative thermal resistance and its influence on microscopic bubble growth. *Int. J. Heat Mass Transf.* **2016**, *101*, 733–741. [CrossRef]
43. Chen, Y.; Jin, S.; Yu, B.; Ling, K.; Sun, D.; Zhang, W.; Jiao, K.; Tao, W. Modeling and study of microlayer effects on flow boiling in a mini-channel. *Int. J. Heat Mass Transf.* **2023**, *208*, 124039. [CrossRef]
44. Kossolapov, A.; Phillips, B.; Bucci, M. Can LED lights replace lasers for detailed investigations of boiling phenomena? *Int. J. Multiph. Flow* **2021**, *135*, 103522. [CrossRef]

energies

MDPI

Article

Thermal Performance Analysis of Micro Pin Fin Heat Sinks under Different Flow Conditions

Jéssica Martha Nunes [1], Jeferson Diehl de Oliveira [2], Jacqueline Biancon Copetti [3], Sameer Sheshrao Gajghate [4], Utsab Banerjee [5], Sushanta K. Mitra [5] and Elaine Maria Cardoso [1,*]

1 Post-Graduation Program in Mechanical Engineering, School of Engineering, UNESP—São Paulo State University, Av. Brasil, 56, Ilha Solteira 15385-000, SP, Brazil
2 Department of Mechanical Engineering, FSG—University Center, Os Dezoito do Forte, 2366, Caxias do Sul 95020-472, RS, Brazil
3 LETEF—Laboratory of Thermal and Fluid Dynamic Studies, UNISINOS—University of Vale do Rio dos Sinos, São Leopoldo 93022-750, RS, Brazil
4 Mechanical Engineering Department, G H Raisoni College of Engineering & Management, Pune 412207, Maharashtra, India
5 Micro & Nano-Scale Transport Laboratory, Department of Mechanical and Mechatronics Engineering, Waterloo Institute for Nanotechnology, University of Waterloo, Waterloo, ON N2L 3G1, Canada
* Correspondence: elaine.cardoso@unesp.br

Abstract: Due to microscale effects, the segmented microchannels or micro pin fin heat sinks emerged as a high thermal management solution. In this context, the present work analyzes the influence of different heights of square micro pin fins with an aligned array and investigates their influence on pressure drop and heat transfer behavior. The HFE-7100 is used as the working fluid, and the pressure drop and surface temperature behavior are analyzed for different mass fluxes and inlet subcooling. The single-phase flow was analyzed numerically using the computational fluid dynamics (CFD) software ANSYS FLUENT® for comparing the simulation results with the experimental data, showing that the highest micro pin fins configuration provides a more uniform and lowest wall temperature distribution compared to the lowest configuration. There is a good agreement between the experimental results and the numerical analysis, with a mean absolute error of 6% for all the considered parameters. For the two-phase flow condition, experimental tests were performed, and for the highest subcooling, an increase in mass flux causes an enhancement in the heat transfer for low heat flux; by increasing heat flux, there is a gradual predominance of boiling heat transfer over convection as the heat transfer mechanism. The pressure drop drastically increases with the vapor amount flowing into the system, regardless of the pin fin height; the boiling curves for the higher fin height show a much smaller slope and a smaller wall superheat than the fin with the smallest height, and consequently, a high heat transfer performance. A larger region of the heat sink is filled with vapor for lower inlet subcooling temperatures, degrading the heat transfer performance compared to higher inlet subcooling temperatures.

Keywords: convective boiling; two-phase flow; pin fin geometry; heat transfer coefficient; pressure drop

Citation: Nunes, J.M.; de Oliveira, J.D.; Copetti, J.B.; Gajghate, S.S.; Banerjee, U.; Mitra, S.K.; Cardoso, E.M. Thermal Performance Analysis of Micro Pin Fin Heat Sinks under Different Flow Conditions. *Energies* **2023**, *16*, 3175. https://doi.org/10.3390/en16073175

Academic Editors: Kyung Chun Kim, Silvia Ravelli and Vladimir Serdyukov

Received: 22 February 2023
Revised: 21 March 2023
Accepted: 27 March 2023
Published: 31 March 2023

1. Introduction

An alternative to modifying the configuration of microchannels to minimize the instabilities presented in this system is using segmented or microfinned microchannels (micro pin fins). Such compact heat sinks can be used in high-power laser cooling systems, high-concentration photovoltaic cell cooling, microreactor cooling, fuel cells, and microchips. In addition to the advantage of segmented microchannels in terms of increasing the heat exchange area to volume ratio, which provides an ability to dissipate higher heat rates, they can also be manufactured on the chip scales of electronic devices. The recent development of microfabrication techniques has allowed complex geometries on a reduced scale; thus, in recent years, several studies involving heat transfer in micro pin-fin heat sinks have been carried out in order to characterize the heat transfer mechanisms and to predict the thermal and fluid dynamic behavior of these systems. The micro fins can have different shapes and sizes and be arranged in different patterns to improve heat transfer [1–3]. It is also noteworthy that the ideal spacing of the fins depends on the working fluid and its subcooling in the system. The mini and microfinned channel arrangements are considered promising structures for compact heat sinks [4].

Deng et al. [5] proposed a new type of heat sink with pin fin-interconnected reentrant microchannels (PFIRM) and tested it in convective boiling, using deionized (DI) water and ethanol as working fluids. Several tests were performed under different subcooling conditions (40 and 10 °C) and mass fluxes (125 to 300 kg/m^{2}s). An increase of 39–284% was observed in the heat transfer coefficient (HTC) for water and 29–220% for ethanol compared to parallel microchannels. Authors attributed this enhancement to the interconnected microchannels, which provide different paths for the vapor bubbles reducing the confinement effect. In addition, the interconnected spaces provided ideal conditions for the nucleation of vapor bubbles, contributing to the heat transfer improvement for the PFIRM. For pressure drop, Deng et al. [5] reported an increase with increasing heat flux and vapor quality; moreover, the mass flux strongly influenced pressure drop at moderated and high heat fluxes.

Recently, Asrar et al. [6] conducted an experimental investigation of convective boiling using R245fa in micro gaps improved with the arrays of micro pin fins made of silicon (cylindrical pin fins 150 μm in diameter and 200 μm of interfin space in a staggered arrangement). Several tests were performed at different mass flux conditions (between 781 and 5210 kg/m^{2}s) and inlet temperatures (between 13 and 18 °C). The authors reported that HTC increased with increasing mass flux for the single-phase flow regime. For the two-phase flow regime, they compared the results with their previous works [7]; the new device showed better thermal performance than the previous one. Regarding pressure drop and vapor quality, Asrar et al. [6] found the same behavior as Woodcock et al. [8] and Chien et al. [9], in which these parameters were independent of the heat flux in the single-phase regime but increased remarkably with the intensification of convective boiling.

The scientific community has extensively studied heat transfer in segmented microchannels. However, there are still several challenges that require further research to understand and optimize the process, such as (i) the good balance between heat transfer and pressure drop to have an efficient heat sink, (ii) the appropriate material for the heat sink and the micro pin fins in order to have high thermal conductivity and good mechanical strength, and (iii) the optimal configuration for a micro pin fin heat sink to improve the heat transfer capacity.

Many of the developed works take into account different dimensions and configurations of micro pin fins to understand the physical mechanisms responsible for heat transfer enhancement in an attempt to develop models to be applied on an industrial scale capable of predicting the heat transfer coefficient, the behavior of the critical heat flux, and the pressure drop. As mentioned by Jung et al. [10], since water is more widely available and has superior thermophysical properties, most studies are conducted with it; however, its high electrical conductivity limits its use for embedded cooling.

In this context, the current work investigates the thermal performance and pressure drop of HFE-7100 in micro pin fin heat sinks with different heights and their influence on pressure drop and heat transfer behavior. The effects of geometrical parameters and operating conditions on the thermal performance and pressure drop were analyzed experimentally and, for single-phase flow, numerically. The computational model used in the current study validates the thermal performance and pressure drop determined from the experiments for single-phase flow. The current study contributes to better comprehending heat removal capability, factors impacting heat transfer performance, and mechanisms responsible for enhancing heat transfer in such compact heat exchangers.

2. Materials and Methods

2.1. Experimental Apparatus

Figure 1 shows the experimental apparatus used in the present study. The working fluid is pumped from a reservoir to the flow loop; the HFE-7100 flow rate is set by a Coriolis mass flow meter (Yokogawa ROTAMASS Total Insight with 0.2% mass flow accuracy) installed just upstream of the preheater (consisting of a horizontal copper tube heated by an electrical tape resistance). There is a bypass line used for the test facility maintenance. The pressure drop between inlet and outlet plenums is measured by two pressure transducers (OMEGA PX309 model, with 0.25% accuracy). The flow temperature is measured using previously calibrated K-type thermocouples (uncertainty of 0.3 °C) in the inlet and outlet plenums (both in contact with the fluid). The working fluid is cooled by a condenser and then returned to the reservoir, as shown in Figure 1.

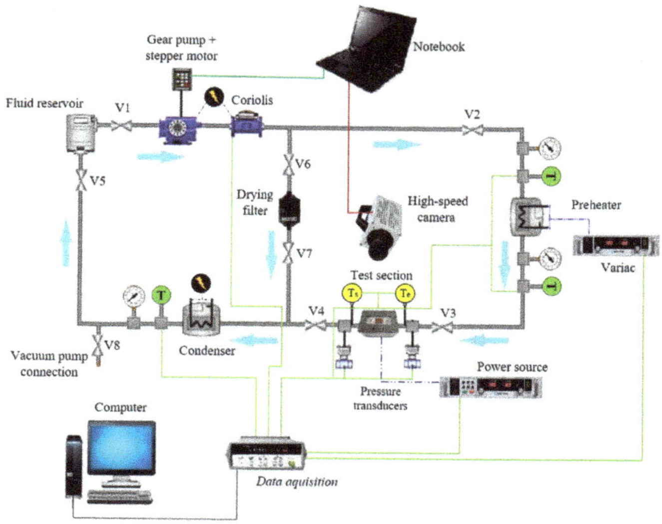

Figure 1. Schematic diagram of the experimental apparatus.

The design of the heat sink test section in different views is shown in Figure 2, including a cutaway to view internal details and elemental descriptions (Figure 2b). Five holes (1 mm diameter) were drilled in the copper block to accommodate K-type thermocouples (Figure 2a, A) to determine the wall temperature and verify the one-dimensional heat conduction along the copper block. The heat flux is provided by electrical resistance (cartridge type, 250 W/220 V) embedded in the copper block (Figure 2a, B) and controlled by a DC power supply.

The one-dimensional conduction law of Fourier was determined with the temperature readings from the thermocouples fixed along the vertical direction of the copper block; therefore, the wall temperature was obtained by extrapolating the linear temperature profile, which exhibits an R-square error of approximately 1.0 regardless of heat flux (Figure 2c). Moreover, the heat losses were lower than 15%, corroborating the one-dimensional steady-state heat flux assumption.

The heat sink (Figure 2) consists of a copper block with a 20 × 15 mm footprint with 972 micro pin fins. The square micro pin fins (300 μm in width and 250 μm of interfin space) were manufactured using a CNC precision milling machine (Hermile, model C800U). The micro fins were manufactured in an aligned array with the fluid flow direction (Figure 3), and different heights (H) of micro pin fins—160 μm named S1 and 350 μm named S2—were analyzed in the current work.

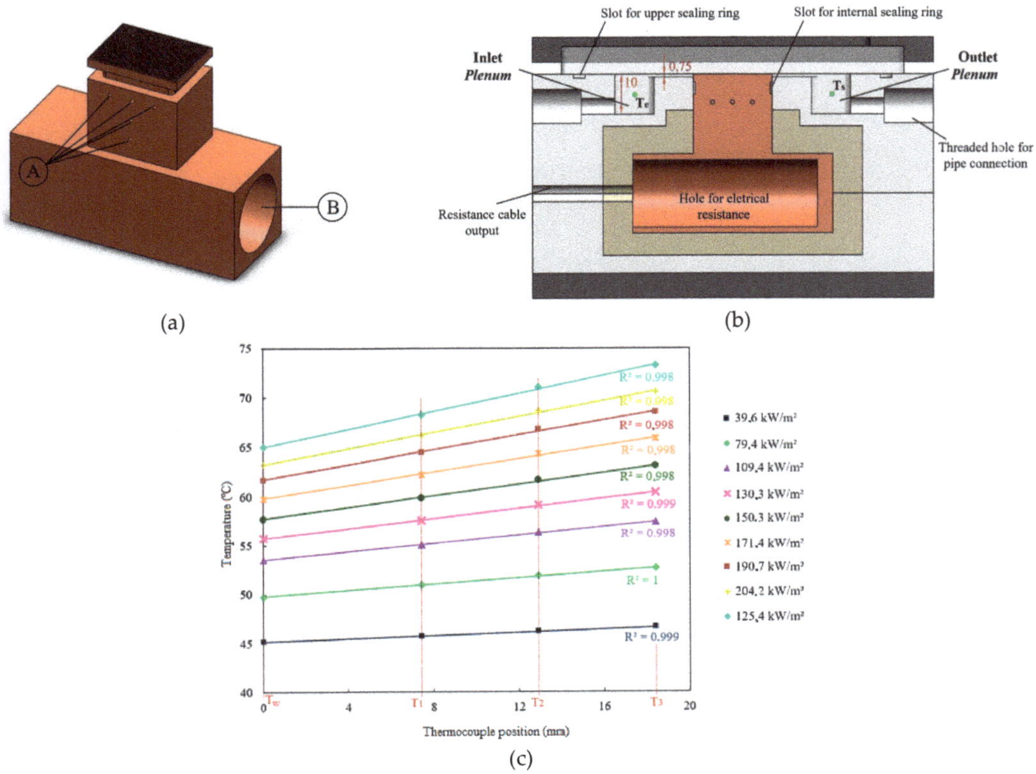

Figure 2. Design of the microfinned heat sink. (**a**) Isometric view; (**b**) front view with internal details (measurements in mm); (**c**) linear temperature profiles used to estimate the wall temperatures.

Figure 3. The constructive array of the micro pin fins.

As shown in Figure 2b, the thermal insulation is made with ceramic and polytetrafluoroethylene (PTFE); the working fluid is not heated before contacting the micro pin fins since the inlet and outlet plenums are machined on the PTFE with 10 × 15 × 10 mm. Two K-type thermocouples, one in the inlet and another in the outlet, measure the working fluid temperature (T_i and T_o, respectively). Flow homogenization channels with a depth of 0.75 mm were manufactured between the plenums and the heat sink to minimize flow entrance turbulence. Flow visualization (using a high-speed camera Photron SA3 model with 1000 fps and 1024 × 1024 resolution) is allowed by a polycarbonate plate (8 mm thick) covering the heat sink.

The geometric characterization of the micro pin fin heat sinks was performed by Zeiss® SteREO DiscoveryV8 and SEM EVO LS15 Zeiss® (Table 1).

Table 1. Structural characterization of the micro pin fin heat sinks.

Surface	STEREO Top View	STEREO Side View	SEM (100×)
S1 H = 160 μm			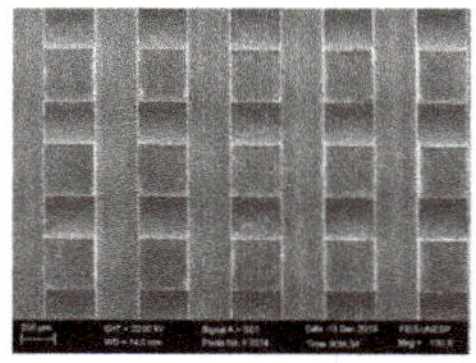
S2 H = 350 μm		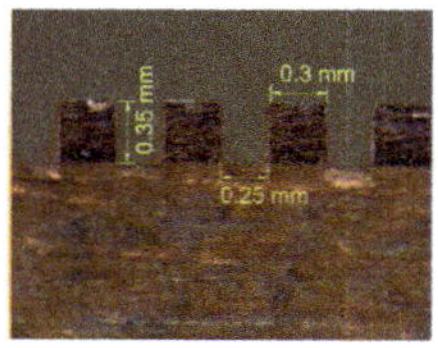	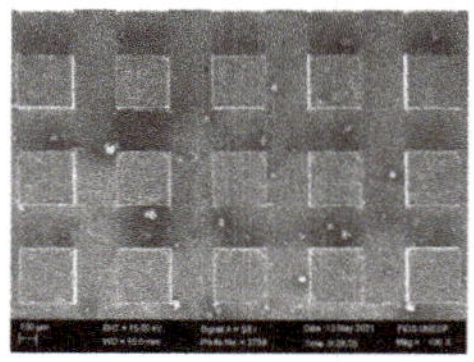

The experimental uncertainty was calculated using the free package developed in Python, called Uncertainties (© 2010–2016, EOL), based on the Taylor series method and standardized by the Bureau International des Poids et Mesures (BIPM). Consequently, for all tests carried out in the current study, the uncertainty of the heat flux, the heat transfer coefficient, and pressure drop varied from 4 to 16%, 7 to 21%, and 3 to 9%, respectively. It is worth mentioning that all analyses take into account the effective heat flux, determined by subtracting the heat loss to the surroundings from the power supplied; in the current study, the heat losses are less than 22% for all tests performed.

2.2. Experimental Procedure

The consistency analysis aims to verify the coherence of the results obtained experimentally; thus, the results for the single-phase flow regime were compared to those obtained from a numerical analysis considering the same conditions. The simulation was based on the mass, momentum, and energy conservation equations with the second-order upwind scheme for energy and pressure and the first-order for momentum [11]. The fluid flow was assumed to be steady-state, incompressible, and laminar, and it was solved by adopting the Finite-Volume Method implemented in ANSYS® Fluent 2020 R2. The computational domain with appropriate boundary conditions is shown in Figure 4.

As a reference pressure, atmospheric pressure was defined; considering the characteristics of a low-pressure system, the outlet pressure was set up as zero. The no-slip condition was considered on all the surfaces. The input parameters for the simulation, such as inlet mass flux and the dimensions of the micro pin fins, were taken from the experimental approach. The heat flux was distributed through the micro pin fins, except for the top side, which was thermally insulated by a polycarbonate piece; for the inlet and outlet plenums, adiabatic wall conditions were considered.

The hex-dominant meshing grid scheme with a free-face mesh type combining triangles and quadrilaterals was used to mesh the systems (for both cases), as shown in Figure 5. The mesh was accomplished in the meshing module with minimum mesh orthogonality of 0.311 (S1, H = 160 μm) and 0.346 (S2, H = 350 μm) and a maximum skewness of 0.56 (S1, H = 160 μm) and 0.61 (S2, H = 350 μm).

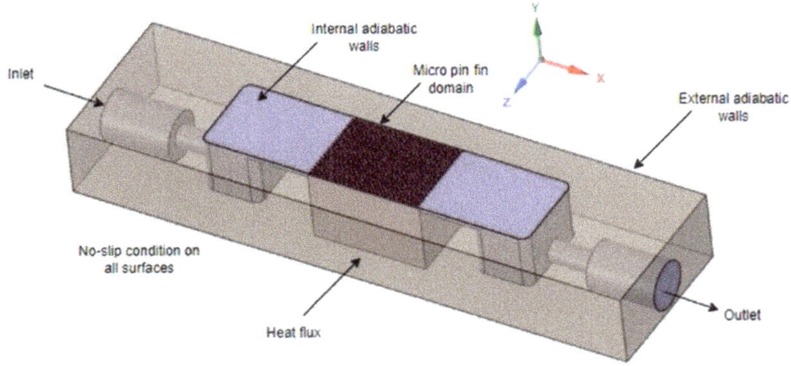

Figure 4. Computational domain of the heat sink with main boundary conditions.

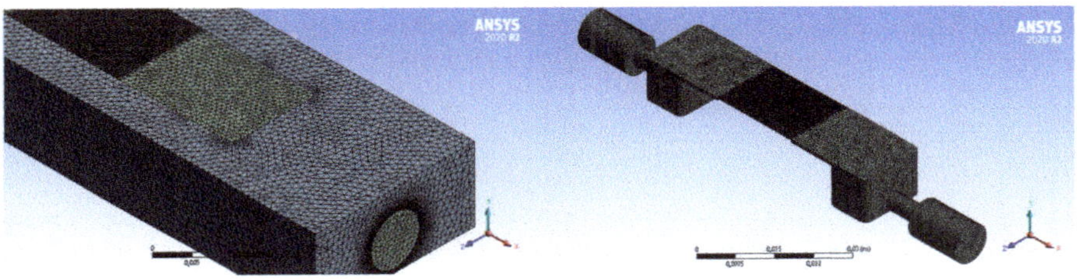

Figure 5. Grid view with fluid domain on the right side.

The convergence occurred for meshes with 511.09 k elements; the finer mesh was achieved when residuals were less than 10^{-5} for the continuity equation and 10^{-6} for momentum and energy equations. The simulations used the segregated algorithm with the SIMPLE algorithm for pressure-velocity coupling.

A study on grid independence was conducted based on wall temperature as a criterion to ensure the results were independent of the mesh. This analysis consisted of three different meshes (around 120 k; 370 k; 550 k elements), aiming to obtain a heated wall temperature range of a maximum of 1.2 K (less than 10% of the maximum wall temperature, according to [12]); it was noticed that the difference in wall temperature was around 1.2 K between the last two meshes. Hence, the mesh chosen aimed to save computational time.

The mean absolute errors ($MAE = \frac{1}{N}\sum_1^N \frac{|\phi_{exp} - \phi_{num}|}{\phi_{exp}} \times 100\%$) of the surface temperature and total pressure drop were 5.6% and 7.1%, respectively. The computational results were consistent with the experimental data for the heat transfer coefficient; for both S1 and S2, the MAE was 5%. Therefore, the mean absolute errors of the experimental and simulation data for pressure drop and heat transfer coefficient are within the experimental uncertainty range.

After the validation analysis for the single-phase flow, two-phase flow tests were performed for two different subcooling values, 10 °C and 20 °C, for mass fluxes of 1000 and 1200 kg/m^{2}s, and for different footprint heat fluxes from 10 kW/m^2 to the system limit, characterized by intense instability in the flow (reverse flow). The gear pump's rotation was set to achieve the desired mass flux; the preheater was adjusted until its outlet temperature was equal to the desired subcooling. A data acquisition system (Agilent 34970A) recorded the data every 2 s after the system achieved the steady-state regime, characterized by temperature variations lower than the thermocouples uncertainties (±0.3 °C). At least 250 data points were recorded, corresponding to 500 s steady-state. The pressure, temperatures, mass flux, and electrical voltage are constantly monitored. Flow visualization was carried out using a high-speed camera. The same procedure is adopted during all the experimental tests to ensure repeatability.

2.3. Data Reduction

The heat transfer coefficient is calculated based on Equations (1)–(5), similar to the approach adopted by Prajapati et al. [13]:

$$\dot{Q}_{loss} = \dot{Q}_{in} - \dot{m} \cdot c_p (T_o - T_i) \quad (1)$$

where $\dot{m}$ corresponds to the mass flow rate [kg/s]; c_p to the specific heat capacity [J/kg·K]; and T_i and T_o are the coolant temperature at the inlet and outlet, respectively. In the current study, heat loss ($\dot{Q}_{loss}$) varied from 15 to 30% over the range of varying parameters. The heat flux, q'', dissipated by the test section is given by:

$$q''_{footprint} = \frac{\dot{Q}_{in} - \dot{Q}_{loss}}{A_p} \quad (2)$$

where A_p is the footprint area of the heating surface. The effective heat flux, q''_{eff} [W/m^2], based on the total surface area in contact with the working fluid (A_t), is calculated by:

$$q''_{eff} = \frac{\dot{Q}_{in} - \dot{Q}_{loss}}{A_t} \quad (3)$$

In order to calculate the total surface area, A_t, the fin parameters and efficiency (η) concepts have been calculated considering the adiabatic fin tip, since a polycarbonate plate is used to cover the heat sink. Thus, A_t is given by Equation (4), where N is the total number of micropillars.

$$A_t = (A_p - N \cdot A_c) + \eta \cdot N \cdot P_{ma} \cdot H \quad (4)$$

where A_c is the cross-sectional area, P_{ma} is the pin fin perimeter, and H is the height of the micro pin fins.

Therefore, it is possible to calculate the heat transfer coefficient or HTC (h) through Equation (5), where T_w is the average temperature of the heat sink provided by three K-type thermocouples fixed within the heat sink wall. The T_f is the average temperature of the fluid given by the same procedure as Leão et al. [14]

$$h = \frac{q''_{eff}}{T_w - T_f} \quad (5)$$

Pressure transducers (at inlet and outlet plenums, P_i and P_o, respectively) measure the pressure drop in the region between the inlet and outlet plenums; thus, the pressure drop through the microchannels is given by $\Delta P = (P_i - P_o) - \Delta P_{contraction} - \Delta P_{expansion}$ where the pressure drop due to contraction and expansion is obtained by the method described in Chalfi and Ghiaasiaan [15].

3. Results and Discussion

3.1. Effect of the Inlet Subcooling Temperature

Figure 6 shows the effect of different inlet subcooling temperatures (10 and 20 °C) on the flow boiling heat transfer for both surfaces (S1 and S2). The increase in subcooling shifted the boiling curve to the left regardless of micro pin fin height. The HTC continuously increased with heat flux for all mass fluxes and higher inlet subcooling temperature, while the HTC slightly decreased with high heat fluxes for lower inlet subcooling temperature. According to Yin et al. [16], such HTC behavior is due to the flow pattern transition into a confined annular flow, where partial dryout occurs on the surface as heat flux increases, leading to the rise in the wall temperature (being more pronounced for S1).

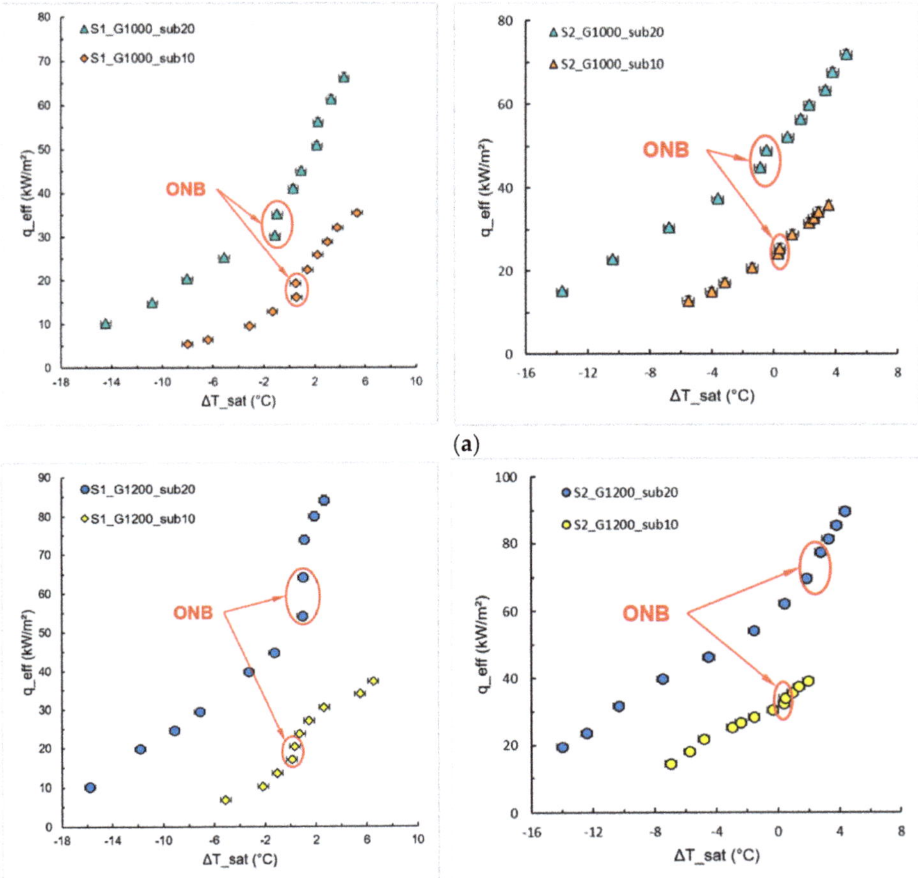

(a)

(b)

Figure 6. Effect of inlet subcooling temperature on flow boiling heat transfer of HFE-7100 for S1 and S2. (**a**) $G = 1000\ kg/m^2s$; (**b**) $G = 1200\ kg/m^2s$.

Analyzing the boiling curves of samples S1 and S2 is possible to observe the beginning of the nucleate boiling regime, indicated in Figure 6 as ONB (Onset of Nucleate Boiling), and characterized by the sudden change in the slope of the boiling curve, reducing the surface temperature.

Figure 7 shows the effect of different inlet subcooling temperatures (10 and 20 °C) on the pressure drops for both surfaces (S1 and S2). One can observe that the effect of inlet subcooling temperatures on both surfaces' pressure drops in the single-phase flow region (for $q'' < 30\ kW/m^2$) is negligible. However, in the case of the two-phase flow region, the pressure drop becomes more significant as the inlet subcooling temperature decreases, regardless of the mass flux and micro pin fin height; a lower inlet subcooling temperature leads to a higher vapor quality through the heat sink, which increases the pressure drop.

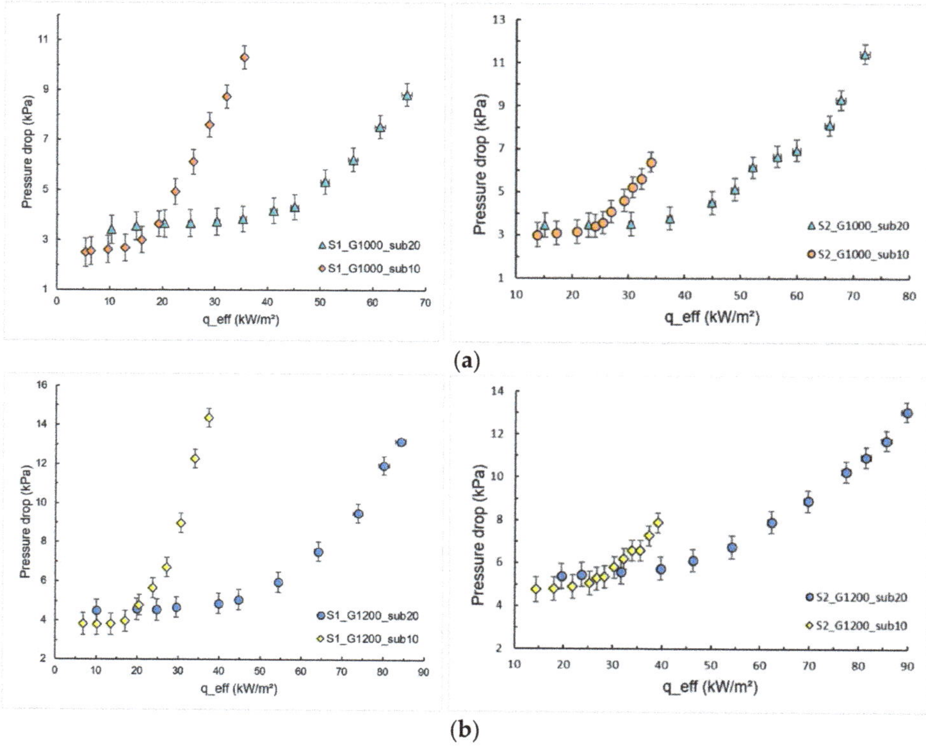

Figure 7. Effect of inlet subcooling on the pressure drops for S1 and S2. (**a**) G = 1000 kg/m^{2}s; (**b**) G = 1200 kg/m^{2}s.

3.2. Effect of the Mass Flux

Figure 8 shows the effect of different mass fluxes on the boiling curves for S1 and S2, with different subcooling temperatures at the inlet of the heat sink. The influence of mass flux, G, on the convective flow boiling heat transfer was negligible for the inlet subcooling of 10 °C and the lowest micro pin fin height (S1). On the contrary, for the highest fin height, the fluid has more space to flow between the fins, and the convective effects (mass flux influences) are more pronounced in the single-phase flow region.

For the inlet subcooling of 20 °C, an increase in the mass flux shifted the curves to the left, characterized by an HTC enhancement. Cheng and Wu [17] indicated a gradual predominance of boiling heat transfer over convection as heat flux increases; furthermore, the micro pin fins induced flow turbulence and strengthened convection heat transfer, the primary heat dissipation component in subcooled convective boiling [18].

For both S1 and S2, increasing the mass flux increased the pressure drops for low heat flux values (single-phase flow region); however, no significant influence of mass flux on pressure drop was observed in the single-phase flow region for both inlet subcooling temperatures (10 and 20 °C). As the heat flux increased (two-phase flow region), the pressure drop became more pronounced due to the vapor mass flowing through the heat sink; thus, the pressure drop was more influenced by the void fraction than by mass flux.

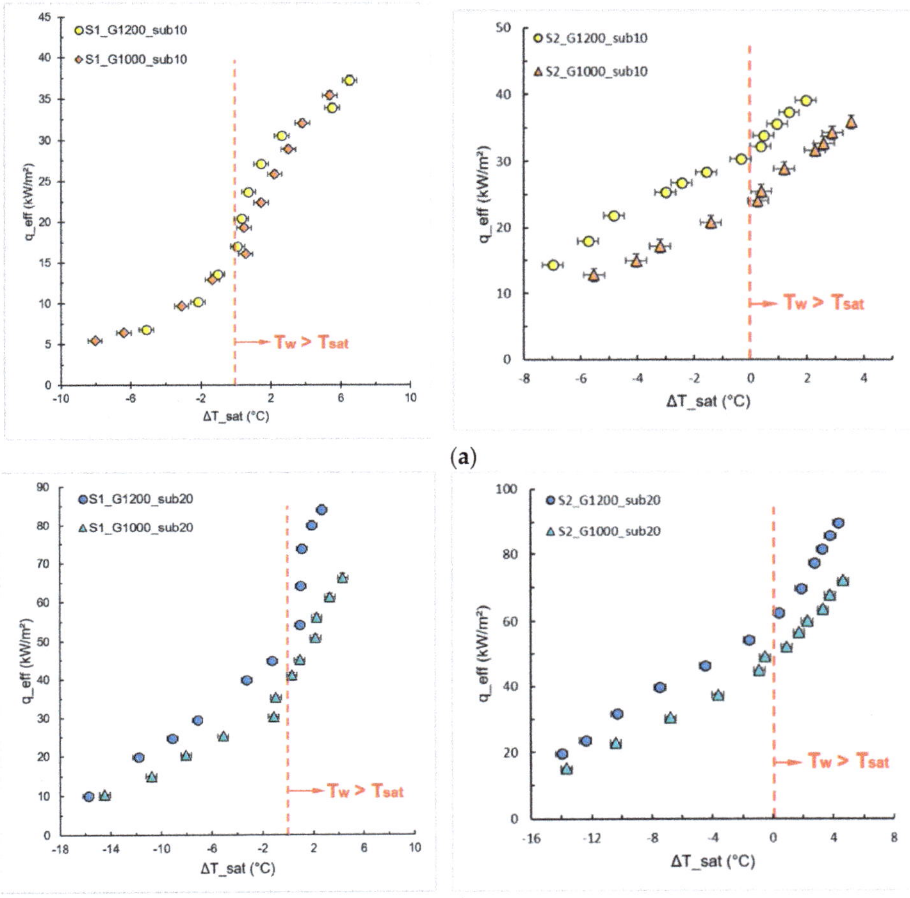

Figure 8. Effects of mass flux on the flow boiling heat transfer of HFE-7100. (**a**) ΔT_{sub} = 10 °C; (**b**) ΔT_{sub} = 20 °C.

3.3. Effect of the Fin Height

Figure 9 shows the effect of pin fin height on the boiling curves for different mass fluxes and a subcooling inlet temperature of 10 °C. Considering the effective heat exchange area, we can infer that the increase in the effective area leads to an increase in the HTC, characterized by the shift of the boiling curve to the left. The same was reported by Kiyomura et al. [19], who evaluated different configurations of micro fin surfaces during pool boiling of the HFE-7100. One can observe in Figure 9 that S2 presents a better HTC, since its effective heat exchange area is approximately 55% greater than S1.

In order to discuss the flow boiling behavior, Figure 10 presents the boiling curve for S1 with G = 1200 kg/m^{2}s and subcooling of 20 °C, with the respective visualization points. It is worth mentioning that similar behavior was observed for all test conditions. Flow boiling videos under these conditions can be found in the Supplementary Material.

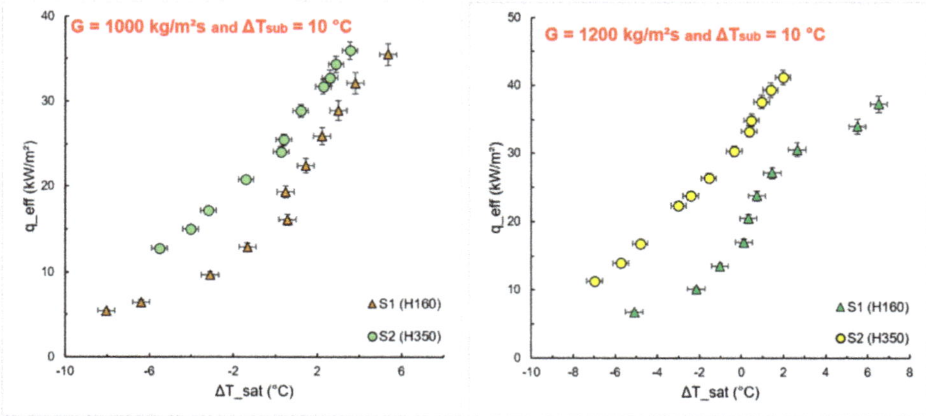

(**a**) (**b**)

Figure 9. Effect of pin fin height on flow boiling heat transfer of HFE-7100 for ΔT_{sub} = 10 °C. (**a**) G = 1000 kg/m^{2}s; (**b**) G = 1200 kg/m^{2}s.

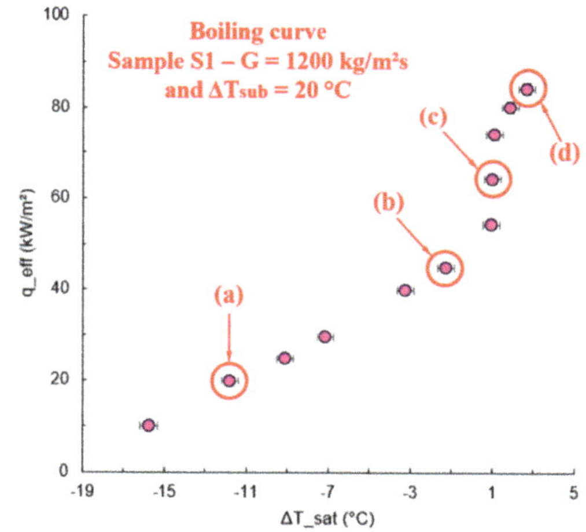

Figure 10. Boiling curve and high-speed camera images for S1. G = 1200 kg/m^{2}s and ΔT_{sub} = 20 °C.

Initially, the single-phase flow regime is predominant at lower heat flux, with no vapor bubbles (point (a), Figure 10). By increasing heat flux, isolated vapor bubbles nucleate preferentially between the adjacent fins, even though the working fluid temperature is lower than the saturation temperature, i.e., subcooled boiling condition (point (b), Figure 10). In the nucleate boiling region, after the ONB, nucleation sites are activated over the entire heating surface (point (c), Figure 10), increasing the departure frequency and the coalescence of vapor bubbles near the heat sink outlet. For high heat fluxes, the vapor core fills the entire length of the heat sink (point (d), Figure 10), and the annular flow regime becomes pronounced. A high void fraction is observed at the outlet of the heat sink, promoting thermal instabilities, a high pressure drop and the occurrence of reverse flow, which is mainly observed for lower inlet subcooling temperature (reverse flow visualization).

4. Conclusions

The current work experimentally studied the thermal and fluid dynamic behaviors, in terms of heat transfer coefficient and pressure drop, of convective boiling using HFE-7100 as the working fluid in a heat sink based on square micro pin fins. Different micro pin fins were tested (heights of 160 and 350 μm in an aligned array) at different mass fluxes (1000 and 1200 kg/m^{2}s) and two levels of inlet subcooling temperatures (10 and 20 °C). The boiling heat transfer and pressure drop behaviors were evaluated for each test condition. The visualization of the experimental tests was performed using a high-speed camera to observe the transition from single-phase to two-phase flow and to identify possible flow patterns and the occurrence of reverse flow. The main conclusions are summarized below:

- ✓ As the mass flux increases, HTC increases in the region where the effects of forced convection are dominant for each sample. However, when the effects of nucleate boiling overlap, the increase in mass flux does not guarantee a gain in HTC, especially for aligned arrays.
- ✓ The lower the inlet subcooling temperature, the lower the heat flux for the ONB occurrence, and a larger region of the heat sink is filled with vapor, which can promote the dryout incipience (decreasing the maximum heat flux).
- ✓ With a lower mass flux and inlet subcooling, the system becomes more sensitive to the effects of nucleate boiling, with significant gains in HTC due to the phase-change heat transfer (for S1 with G = 1000 kg/m^{2}s and ΔT_{sub} = 10 °C, the HTC was increased about 39% compared to ΔT_{sub} = 20 °C for a heat flux of 30 kW/m^2). However, this can lead to the early dryout process.
- ✓ Pressure drop increases substantially with an increase of vapor amount flowing into the heat sink, which becomes more pronounced for lower subcooling, leading to the fluid dynamic limit of the system at lower heat fluxes compared to higher subcooling.
- ✓ An increase in the effective area leads to an increase in the HTC; thus, the taller the micro pin fins, the higher the heat exchange area, leading to an HTC enhancement.
- ✓ The reverse flow occurrence was observed more intensely for the lowest inlet subcooling temperature; the high vapor core acts as a barrier to the flow, degrading the HTC, increasing the pressure drop, and causing thermal and fluid dynamic instabilities.

This study indicates that further attention must be given concerning physical parameters (related to surface and working fluid) to the development of new technologies for thermal management systems. An isolated analysis of the effects of surface characteristics or flow parameters is not sufficient to explain the HTC and pressure drop behavior. The optimal configuration for a micro pin fin heat sink will depend on several factors, including the heat transfer requirements, the fabrication process, and the fluid flow properties. More analyses can be conducted in future works, such as developing new correlations to evaluate the HTC; for that, an extensive experimental database is needed considering different design configurations and operating conditions.

Supplementary Materials: The following supporting information can be downloaded at https://www.mdpi.com/article/10.3390/en16073175/s1: Video S1: point (a), Figure 10; Video S2: point (b), Figure 10; Video S3: point (c), Figure 10; Video S4: point (d), Figure 10; Video S5: reverse flow visualization.

Author Contributions: All authors contributed equally to developing the manuscript. All authors have read and agreed to the published version of the manuscript.

Funding: This research was funded by Conselho Nacional de Desenvolvimento Científico e Tecnológico (CNPq), grants numbers 458702/2014-5 and 309848/2020-2, and Fundação de Amparo à Pesquisa do Estado de São Paulo (FAPESP), grants numbers 2013/15431-7, 2019/02566-8, and 2022/03946-1.

Data Availability Statement: The data supporting this study's findings are available upon request.

Acknowledgments: The authors are grateful for the financial support from the UNESP, CAPES, CNPq, and FAPESP. The authors also thank Alessandro Roger Rodrigues from Escola de Engenharia de São Carlos/USP and Ricardo Arai from IFSP/São Carlos for their important contributions to this work.

Conflicts of Interest: The authors declare no conflict of interest, and the funders had no role in the study's design; in the collection, analyses, or interpretation of data; in the writing of the manuscript; or in the decision to publish the results.

References

1. Tullius, J.; Tullius, T.; Bayazitoglu, Y. Optimization of short micro pin fins in minichannels. *Int. J. Heat Mass Transf.* **2012**, *55*, 3921–3932. [CrossRef]
2. Liang, G.; Mudawar, I. Review of pool boiling enhancement by surface modification. *Int. J. Heat Mass Transf.* **2019**, *128*, 892–933. [CrossRef]
3. Li, W.; Dai, R.; Zeng, M.; Wang, Q. Review of two types of surface modification on pool boiling enhancement: Passive and active. *Renew. Sustain. Energy Rev.* **2020**, *130*, 109926. [CrossRef]
4. McNeil, D.A.; Raeisi, A.H.; Kew, P.A.; Hamed, R.S. An investigation into flow boiling heat transfer and pressure drop in a pin–finned heat sink. *Int. J. Multiph. Flow* **2014**, *67*, 65–84. [CrossRef]
5. Deng, D.; Chen, L.; Wan, W.; Fu, T.; Huang, X. Flow boiling performance in pin fin-interconnected reentrant microchannels heat sink in different operational conditions. *Appl. Therm. Eng.* **2019**, *150*, 1260–1272. [CrossRef]
6. Asrar, P.; Ghiaasiaan, S.M.; Joshi, Y.K. Two-Phase Heat Transfer and Flow Regimes in Pin Fin-Enhanced Microgaps—Effect of Pin Spacing. *ASME J. Heat Transf.* **2021**, *143*, 023001. [CrossRef]
7. Asrar, P.; Zhang, X.; Green, C.E.; Bakir, M.; Joshi, Y.K. Flow boiling of R245fa in a microgap with staggered circular cylindrical pin fins. *Int. J. Heat Mass Transf.* **2018**, *121*, 329–342. [CrossRef]
8. Woodcock, C.; Yu, X.; Plawsky, J.; Peles, Y. Piranha Pin Fin (PPF)—Advanced flow boiling microstructures with low surface tension dielectric fluids. *Int. J. Heat Mass Transf.* **2015**, *90*, 591–604. [CrossRef]
9. Chien, L.H.; Cheng, Y.T.; Lai, Y.L.; Yan, W.M.; Ghalambaz, M. Experimental and numerical study on convective boiling in a staggered array of micro pin-fin microgap. *Int. J. Heat Mass Transf.* **2020**, *149*, 119203. [CrossRef]
10. Jung, D.; Lee, H.; Kong, D.; Cho, E.; Jung, K.W.; Kharangate, C.R.; Iyengar, M.; Malone, C.; Asheghi, M.; Lee, H.; et al. Thermal design and management of micro-pin fin heat sinks for energy-efficient three-dimensional stacked integrated circuits. *Int. J. Heat Mass Transf.* **2021**, *175*, 121192. [CrossRef]
11. Ortegon, J.A.A.; Souza, R.R.; Silva, J.B.C.; Cardoso, E.M. Analytical, experimental, and numerical analysis of a microchannel cooling system for high-concentration photovoltaic cells. *J. Braz. Soc. Mech. Sci. Eng.* **2019**, *41*, 255. [CrossRef]
12. Computational Fluid Dynamics Committee. *Guide for the Verification and Validation of Computational Fluid Dynamics Simulations (AIAA G-077-1998(2002))*; American Institute of Aeronautics and Astronautics, Inc.: Washington, DC, USA, 1998.
13. Prajapati, Y.K.; Pathak, M.; Khan, M.K. Bubble dynamics and flow boiling characteristics in three different microchannel configurations. *Int. J. Therm. Sci.* **2017**, *112*, 371–382. [CrossRef]
14. Leão, H.L.S.L.; Nascimento, F.J.; Ribatski, G. Flow boiling heat transfer of r407c in a microchannels based heat spreader. *Exp. Therm. Fluid Sci.* **2014**, *59*, 140–151. [CrossRef]
15. Chalfi, T.Y.; Ghiaasiaan, S. Pressure drop caused by flow area changes in capillaries under low flow conditions. *Int. J. Multiph. Flow* **2008**, *34*, 2–12. [CrossRef]
16. Yin, L.; Chauhan, A.; Recinella, A.; Jia, L.; Kandlikar, S.G. Subcooled flow boiling in an expanding microgap with a hybrid microstructured surface. *Int. J. Heat Mass Transf.* **2020**, *151*, 119379. [CrossRef]
17. Cheng, X.; Wu, H. Improved flow boiling performance in high-aspect-ratio interconnected microchannels. *Int. J. Heat Mass Transf.* **2021**, *165*, 120627. [CrossRef]
18. Yin, L.; Jiang, P.; Xu, R.; Hu, H. Water flow boiling in a partially modified microgap with shortened micro pin fins. *Int. J. Heat Mass Transf.* **2020**, *155*, 119819. [CrossRef]
19. Kiyomura, I.S.; Nunes, J.M.; de Souza, R.R.; Gajghate, S.S.; Bhaumik, S.; Cardoso, E.M. Effect of microfin surfaces on boiling heat transfer using HFE-7100 as working fluid. *J. Braz. Soc. Mech. Sci. Eng.* **2020**, *42*, 366. [CrossRef]

energies

Article

Experimental Study of Mass Transfer in a Plug Regime of Immiscible Liquid–Liquid Flow in a T-Shaped Microchannel

Semyon Vostretsov [1,2], Anna Yagodnitsyna [1,2,*], Alexander Kovalev [1,2] and Artur Bilsky [1,*]

1 Kutateladze Institute of Thermophysics SB RAS, 630090 Novosibirsk, Russia
2 Physics Department, Novosibirsk State University, 630090 Novosibirsk, Russia
* Correspondence: yagodnitsinaaa@gmail.com (A.Y.); bilsky@itp.nsc.ru (A.B.)

Abstract: In the presented work, the influence of parameters such as the total flow rate of phases, the ratio of flow rates, and residence time on mass transfer during the two-phase flow of immiscible liquids in a T-shaped microchannel was investigated using the micro-LIF technique. The study focused on the plug flow regime, where a 70% water–glycerol solution was used as the dispersed phase, and tri-n-butyl phosphate (TBP) was used as the carrier phase. We determined the transition boundary between the dispersed and parallel flow patterns and calculated the plug length and velocities to develop a mass transfer model. Furthermore, we measured the partition coefficient for the set of liquids used in the experiments and analyzed the concentration fields inside the slugs of the continuous phase at various distances downstream of the T-junction. Using the obtained data, we determined the extraction efficiency and overall volumetric mass transfer coefficient and established dependencies demonstrating the effect of the flow-rate ratio, total flow rate, and the residence time on mass transfer rate and extraction efficiency. Finally, we developed a model for the overall volumetric mass transfer coefficient corresponding to the set of liquids used with an R-squared value of 0.966.

Keywords: immiscible liquids; extraction; mass transfer; micro-LIF; microchannel

Citation: Vostretsov, S.; Yagodnitsyna, A.; Kovalev, A.; Bilsky, A. Experimental Study of Mass Transfer in a Plug Regime of Immiscible Liquid–Liquid Flow in a T-Shaped Microchannel. *Energies* **2023**, *16*, 4059. https://doi.org/10.3390/en16104059

Academic Editor: Dmitry Eskin

Received: 10 April 2023
Revised: 10 May 2023
Accepted: 11 May 2023
Published: 12 May 2023

1. Introduction

Microchannels are ubiquitous in various fields under their unique properties. Owing to their small size, microchannels exhibit a high surface-to-volume ratio, which allows for achieving high efficiency in heat and mass transfer processes. Therefore, microchannels have found their application as micro-heat exchangers, micromixers, microextractors [1–5], and integrated systems for screening and sorting biological objects [6]. Standard mass transfer equipment, such as columns and settlers, used in these processes have several significant disadvantages, including high reagent consumption, low process intensity, and potential danger to personnel in the case of leaks. In contrast, microchannels are devoid of these disadvantages and can intensify mass transfer. Therefore, studying the influence of various parameters on mass transfer in microchannels is necessary.

The increasing demand for low-carbon, cost-effective energy sources has led to the intensive development of new power plants worldwide [7,8]. As a result, the amount of spent nuclear fuel is constantly growing, and its regeneration is most acute for radiochemical technology. Typically, uranium(VI) and plutonium(IV) are recovered from nitric acid solutions of spent nuclear fuel through the PUREX process, with mixtures of organic solvents with tributylphosphate (TBP) as an extractant [9]. By using microchannels as reservoirs for extraction, the technology can be improved in terms of the safety and efficiency of the extraction process.

The slug regime is of great interest for extraction in a microchannel with a two-phase flow of immiscible liquids. This is because plugs and slugs provide a greater interfacial area-to-volume ratio. In addition, due to vortices in slugs and plugs, mass transfer intensification occurs both inside the phases and between them [10]. Thus, various models have been developed to assess mass transfer for gas–liquid [11,12] and liquid–liquid [13–16] systems in this flow regime. Based on the results obtained in [17–20], conclusions can be drawn regarding the dependence of mass transfer efficiency in microchannels on the residence time, temperature, and viscosity of the liquids used, as well as the shape of the microchannel. A review conducted by Ganguli and Pandit [21] summarized the effect of microchannel size on mass transfer. It was concluded that reducing the size of the cross section of the microchannel increases the intensity of mass transfer. Kashid et al. [17] investigated the influence of microchannel geometry on mass transfer efficiency and found that the microchannels with obstacles inside could intensify mass transfer due to additional flow circulation.

The influence of temperature on mass transfer in a liquid–liquid system was described by Zhang et al. [18]. The highest mass transfer coefficients corresponded to the highest temperatures studied, however, circulation within the phases decreased. The authors suggest that this effect was caused by the lower viscosity of the continuous phase, which led to a lower shear stress acting on the dispersed phase. The influence of the viscosity of working liquids on mass transfer was studied in [18,19]. Based on experiments using various sets of liquids, it was found that the viscosity of the dispersed phase had a negligible effect on mass transfer, in contrast to the viscosity of the continuous phase.

The effect of residence time on mass transfer in the slug flow regime was investigated by Angeli et al. [20]. The authors concluded that mass transfer mainly occurs at short residence times. At longer times, it becomes noticeable that the extraction efficiency increases more with a higher total flow rate, indicating the influence of the flow structure inside the slugs and plugs on mass transfer. Thus, it is possible to distinguish two main stages of mass transfer in the two-phase flow of immiscible liquids. At short residence times, there is a high solute concentration gradient between phases, so the diffusion coefficient determines the mass transfer. At long residence times, the layer near the interfacial boundary becomes poorer, and the hydrodynamics of the flow becomes decisive to ensure the inflow of solute to the interfacial boundary.

There are different approaches to assessing mass transfer in microchannels which can be classified as local and integral techniques. In local measurements, the values characterizing the mass transfer are determined based on concentration fields obtained by the microresolution laser-induced fluorescence technique (micro-LIF) or colorimetric techniques. In micro-LIF measurements, the transferred solute itself is used as a fluorescent dye [20,22]. In colorimetric techniques, the dye added to the flow changes color due to the reaction with the transferred solute or pH change occurring when the compound in the flow reacts with the solute [14,19,23]. However, integral measurements of the solute concentration in phases are often performed. For example, gas chromatography was used by Kashid et al. [17]. After phase separation, chromatography of the examined liquid is carried out, and the amount of solute is determined. Another method of integral measurement is UV spectroscopy, performed by Priest et al. [5] but, in this case, it is necessary to separate the phases before analysis. Moreover, integral techniques are not applicable if the residence time in microchannels is much lower than the phase separation time.

Currently, existing studies on mass transfer in the flow of immiscible liquids in microchannels aim to determine the influence of parameters such as microchannel geometry, flow pattern, total flow rate, the flow rate ratio of the continuous and dispersed phases, and residence time. However, mass transfer studies using local measurement techniques are limited to a narrow range of flow rates and phase flow ratios. The present work uses a micro-LIF technique to study the effect of flow parameters and residence time on the efficiency of mass transfer in the slug flow regime of immiscible liquids in a T-shaped microchannel.

2. Materials and Methods

2.1. Mass Transfer Characteristics

Extraction is a process of transferring a solute between two non-miscible media. In our experiment, the dye is extracted from the dispersed phase into the continuous phase, so the equations presented in this article refer to the slugs of the continuous phase. The main parameters studied during mass transfer are the overall volumetric mass transfer coefficient (1) and extraction efficiency (2):

$$k_L a = \frac{\ln\left(\frac{C_c^{in} - C_c^*}{C_c^{out} - C_c^*}\right)}{\tau}, \quad (1)$$

$$\%E = \frac{C_c^{out}}{C_c^*}, \quad (2)$$

where C_c^{in} and C_c^{out}—the solute concentration in the continuous phase at the microchannel inlet and outlet, respectively, C_c^*—equilibrium concentration, and τ—residence time.

The ratio of equilibrium concentrations of the solute in the phases is called partition coefficient K:

$$K = \frac{C_{org}^*}{C_{aq}^*} = \frac{C_c^*}{C_d^*} \quad (3)$$

where subscripts *org* and *aq* denote organic and aqueous phases.

The equilibrium concentration for various flow rates is determined from the following considerations. Initially, the solute is presented only in the dispersed phase. Its amount N_{in} entering the channel is determined by the flow rate of the dispersed phase Q_d and the initial concentration of the solute in it C_0:

$$N_{in} = Q_d \cdot C_0 \quad (4)$$

At the outlet of the channel, the solute is presented in both phases in concentrations C_c and C_d. Due to the conservation of the amount of solute at the inlet and outlet of the microchannel, it is possible to write:

$$Q_d \cdot C_0 = Q_d \cdot C_d + Q_c \cdot C_c \quad (5)$$

Assuming that equilibrium is established between the phases in the microchannel with respect to mass transfer, we substitute the equilibrium concentrations and express the equilibrium concentration in the continuous phase:

$$C_c^* = \frac{C_0}{\left(\frac{1}{K} + \frac{Q_c}{Q_d}\right)} \quad (6)$$

Thus, to obtain the extraction efficiency and the overall volumetric mass transfer coefficient, it is necessary to have values of the partition coefficient K, the concentration of the solute in the inlet and outlet of the measurement area of the microchannel C^{in} and C^{out}, and the residence time τ.

2.2. Experimental Setup and Techniques

The mass transfer was studied in a plug regime of immiscible liquids in a T-shaped microchannel with a square cross-section of 370 µm × 200 µm. The schemes of the experimental setup and the microchannel are presented in Figure 1. The experimental setup consisted of a Carl Zeiss Axio Observer.Z1 microscope with a lens magnification of M = 10× and a numerical aperture of NA = 0.25. A T-shaped microchannel was mounted on the microscope stage. Liquid flow rates were set using a KDS Gemini 88 double syringe pump with an accuracy of 0.3%. The experiments were performed at room temperature T = 23 °C.

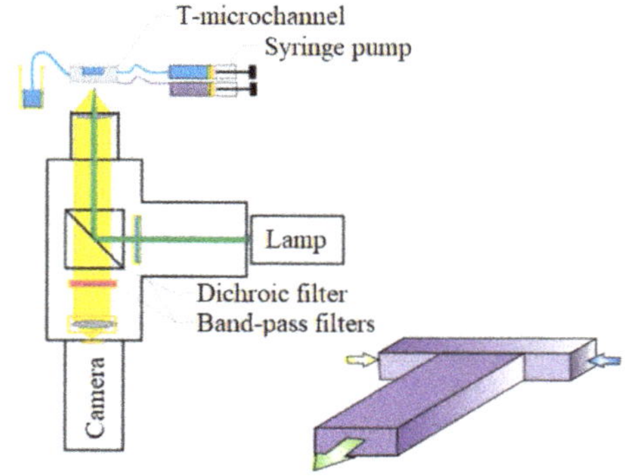

Figure 1. The schemes of the experimental setup and a T-shaped microchannel.

The microresolution laser-induced fluorescence technique was applied to measure the solute concentration fields. The method is based on the relation between dye concentration in the fluid and its fluorescence intensity. In the standard LIF method at the macroscale, the flow is illuminated with a laser sheet, hence the concentration field is determined by the emission intensity of the fluorescent dye in the measurement plane. However, creating a thin laser sheet is impossible on the microscale; therefore, the entire flow volume was illuminated by a mercury lamp, obtaining the concentration field averaged over the channel depth. Rhodamine 6G fluorescent dye was used as a solute added initially to the dispersed phase. The solute concentration measurements were performed in the slugs of the continuous phase since the measurements in the plugs of the dispersed phase are unreliable due to the presence of the film of the continuous phase on the bottom and top walls of the microchannel. The light from the mercury lamp passed through a band-pass filter of 546 $\pm$ 12 nm, reflected from a dichroic mirror with 560 nm edge wavelength, and illuminated the measurement area of the microchannel. The re-emitted light from the solute passed through the dichroic mirror and a band-pass filter of 575–640 nm. Fluorescence intensity was recorded by an IMPERX CCD camera with a resolution of 4 MPix in a 16-bit format. The spatial resolution was 0.43 µm/pixel. The fluorescence images were processed in ActualFlow software, and the calculated concentration fields were analyzed with the help of a script written in Python.

The concentration field calculation during the experiment requires a calibration curve to determine the dependence of the registered light intensity on the dye concentration in the flow. Calibration was performed pixel by pixel, so the measurement area retained the same position during the curve construction and the experiment. For calibration, the microchannel was filled with the liquid with known fluorescent dye concentrations including zero concentration. One hundred fluorescence intensity images were registered for each concentration and averaged before proceeding to the calibration step. Additionally, the dark noise of the camera was captured and subtracted from the calibration images and the fluorescence images of the slugs.

To reveal the flow patterns and plug properties, high-speed flow visualization was performed. The flow was illuminated by a halogen lamp. A microscope lens with a magnification of M = 5× and a numerical aperture of NA = 0.12 was used. Flow images were recorded by a high-speed PCO camera with 1 MPix resolution and a frame rate of up to 1 kHz. The spatial resolution was 2.17 µm/pixel.

An aqueous solution was used as the dispersed phase, and tri-n-butyl phosphate (TBP) was used as the continuous phase. The properties of liquids are presented in Table 1. Water–glycerol solution properties were calculated by the parametrization in [24] with adjustments described in [25]. The physical properties of TBP are provided by the supplier. The interfacial tension of the liquids was evaluated using Antonov's rule [26], the calculated value coincides well with the measured value in [27]. The percentage for the water–glycerol solution was chosen so that its refractive index n_d = 1.428 was close to the refractive index of the carrier phase n_c = 1.425. Therethrough, the distortion of the rays of the re-emitted light is avoided, significantly reducing the measurement error.

Table 1. Physical properties of the liquids.

	Water–Glycerol Solution		TBP
density, kg/m^3	1181		973
dynamic viscosity, mPa·s	23		3.4
interfacial tension, mN/m		12.6	

3. Results

3.1. Slug Flow Properties

To distinguish the boundary between parallel and dispersed flow patterns high-speed visualization was performed in the range of superficial velocities of the continuous phase (100 μm/s $\leq U_c \leq$ 6700 μm/s) and dispersed phase (100 μm/s $\leq U_d \leq$ 2200 μm/s). Subsequently, a flow pattern map was created, as shown in Figure 2a. Initially, the flow rate of the continuous phase was kept constant, after which the flow rate of the dispersed phase was increased until a parallel flow pattern was established in the channel.

The dimensionless criterion $We^{\alpha}\cdot Oh^{b}$ is suitable for unifying flow pattern maps. Here, We is the Weber number, which expresses the ratio of inertia forces to interfacial tension forces, and Oh is the Ohnesorge number, which characterizes the properties of liquids in a two-phase flow [28]. In works by Kovalev et al. [29,30], the parameters a = 0.4 and b = 0.6 were proposed to draw a universal flow pattern map. Additionally, the equation for the boundary between the segmented and continuous flow patterns was suggested as $We_d^{0.4}\cdot Oh_d^{0.6} = 0.052\cdot\left(We_c^{0.4}Oh_c^{0.6}\right)^{0.4}$, which describes the transition between flow patterns for liquid–liquid sets with different properties. We compared the boundary between slug and parallel flow in our experiment with the equation by Kovalev et al. in Figure 2b. The boundary is located at the higher values of the $We_d^{0.4}\cdot Oh_d^{0.6}$ which coincides with the findings by Kovalev et al. that the proposed equation works well for pairs of liquids with a viscosity ratio of less than unity. In our case, the viscosity ratio is 6.8.

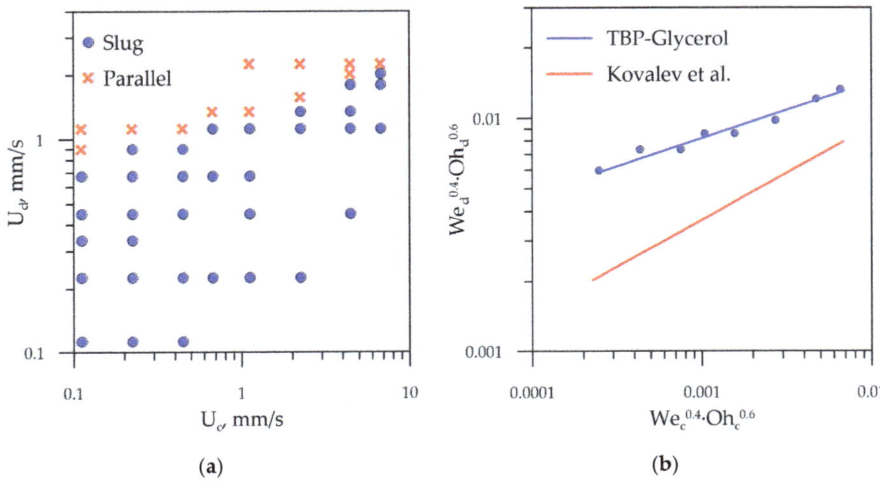

Figure 2. (**a**) Flow pattern map in terms of superficial velocities of the phases; (**b**) the boundaries between parallel and dispersed flow patterns [29,30].

Based on the flow images obtained during high-speed visualization, we measured the lengths and velocities of plugs, which are necessary for further developing the mass transfer model. For each flow rate of the continuous and dispersed phase, the averaging was performed among at least ten slugs. The slug lengths were approximated by models from the works of Garstecki et al. [31] (7) and Xu et al. [32] (8). The first approximation corresponds to the 'squeezing' mechanism of plug breakup, which depends crucially on the blockage of the channel by a liquid plug. Xu's model is designed for the so-called 'dripping' regime of plug formation dominated by the balance between shear force and interfacial force.

$$L_{plug} = 1 + 1.33 \cdot \frac{Q_d}{Q_c} \tag{7}$$

$$L_{plug} = 2.34 * \left(\frac{Q_d}{Q_c}\right)^{0.42} \left(\frac{1}{Ca}\right)^{0.02} \tag{8}$$

where $Ca = (\mu_c U_{bulk})/\sigma$—the capillary number based on the dynamic viscosity of the continuous phase and bulk velocity. The parity plots of the measured plug lengths and approximated data according to Equations (7) and (8) are presented in Figure 3a,b. The R-squared was 0.76 and 0.86 for each of the models, respectively. As we obtained good results for plug approximation by the Xu model, we can conclude that the capillary number of the continuous phase influences the plug length, and the formation of plugs is in the 'dripping' regime.

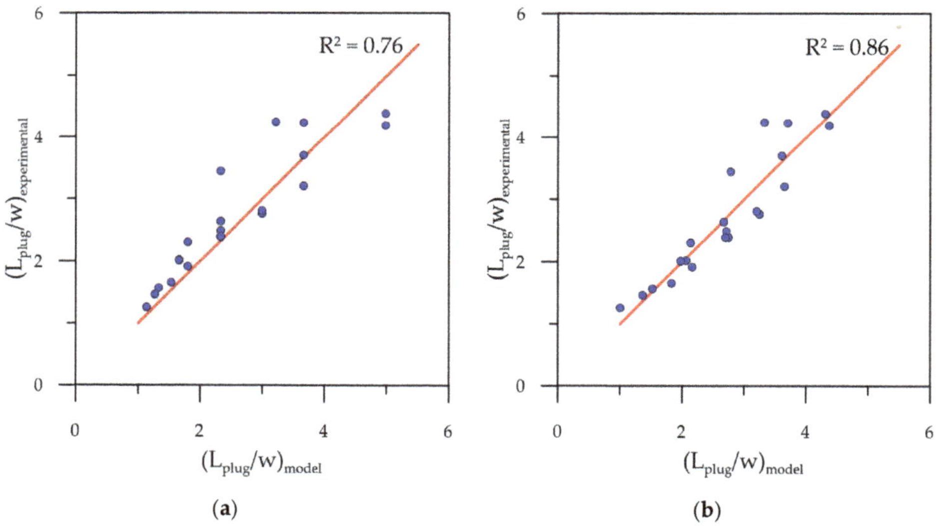

Figure 3. Comparison of experimental plug lengths with models (**a**) Equation (7); (**b**) Equation (8).

The plug velocity was approximated using linear and power functions of bulk velocity. We obtained nearly equal R-squared for each approximation: 0.996 for the linear approximation and 0.997 for a power function. The function plots are presented in Figure 4. Further, we used a linear approximation of the plug length.

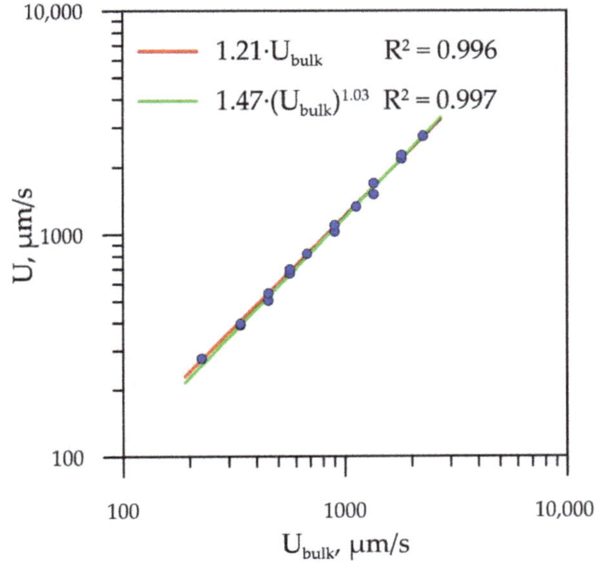

Figure 4. The dependence of plug velocity on bulk velocity.

3.2. Mass-Transfer Assessment

3.2.1. Partition Coefficient Measurement

Three samples were prepared to determine the partition coefficient K. These samples were mixtures of equal volumes of TBP and water–glycerol solution, 30 mL each. The initial concentrations of the fluorescent dye in the dispersed phase were 10 mg/L, 20 mg/L, and 30 mg/L, corresponding to the linear part of the dependence of the emitted light intensity on the dye concentration according to the work of Zehentbauer et al. [33]. Each sample was mixed on a magnetic stirrer for 72 h and allowed to settle until the interfacial layer became completely uniform, indicating complete coalescence of the droplets. The liquids were then manually separated.

The residual concentration of the solute in the aqueous phase was determined using the micro-LIF technique in a microchannel. To construct the calibration curve, the microchannel was filled with prepared samples of the aqueous phase with the known fluorescent-dye concentrations: 0 mg/L, 10 mg/L, 20 mg/L, and 30 mg/L. Based on the obtained intensity values corresponding to specific concentrations, a calibration curve for the aqueous phase was constructed (Figure 5a). The residual concentrations in the separated samples of the aqueous phase were determined using the calibration curve. Since equal volumes of water–glycerol solution and TBP were used, it was possible to calculate the volume concentration of the solute in the TBP samples:

$$C^*_{org} = C_{0\,aq} - C^*_{aq} \tag{9}$$

The partition coefficient was calculated according to Equation (3). Its average value was $K = 3.9 \pm 0.7$.

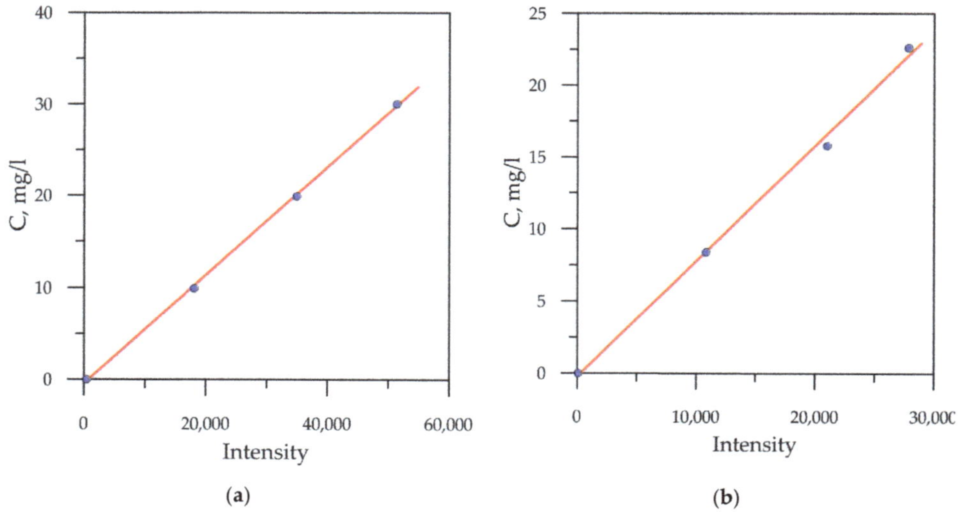

Figure 5. Calibration curves of the registered fluorescent intensity on fluorescent dye concentration: (**a**) water–glycerol solution; (**b**) TBP at first measurement area.

3.2.2. Solute Concentration Measurements in Slugs of the Continuous Phase

To measure solute concentration in slugs of the continuous phase in two-phase flow, the calibration curve of the dependence of Rhodamine 6G fluorescence intensity on its concentrations was derived using TBP samples obtained after phase separation in the measurements of the partition coefficient. The calibration curve is presented in Figure 5b. The maximum deviation of the points from the approximating linear function was 4.6%, and the random measurement error did not exceed 1.2%.

The measurement scheme of solute concentration in TBP in two-phase flow is presented in Figure 6. The continuous phase without a fluorescent dye and the dispersed phase with a Rhodamine 6G concentration of 30 mg/L were fed to the microchannel inlets, which formed slugs and plugs in the mixing zone. The flow rates of the continuous and dispersed phases were chosen according to the plug regime in the flow pattern map obtained during high-speed visualization and the condition that slugs of the continuous phase fit entirely into the CCD camera frame. Thus, the flow rate ranges of the dispersed and continuous phases were 0.5 μL/min $\leq Q_d \leq$ 6 μL/min and 0.5 μL/min $\leq Q_c \leq$ 3 μL/min, and the flow rate ratio Q_d/Q_c varied in the range of 1–6.

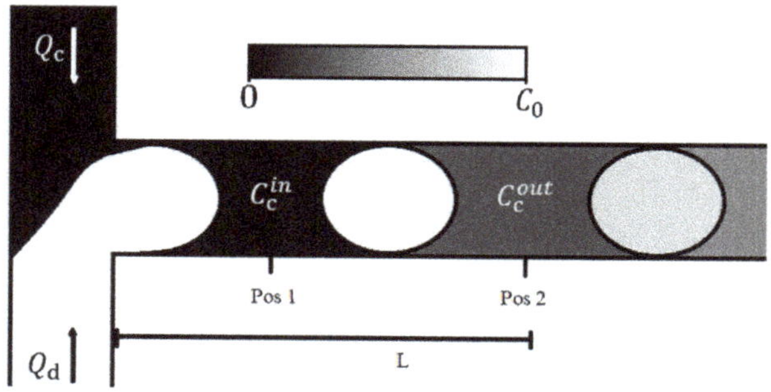

Figure 6. The measurement scheme of the solute concentration in slugs of the continuous phase.

A series of fluorescence images were recorded at points located at distances of 5.6 mm, 8.5 mm, 13.6 mm, and 17 mm, which corresponded to 20, 32, 52, and 65 hydraulic diameters downstream of the T-junction. At least 15 slugs of the continuous phase were recorded for each studied regime and position. Then, concentration fields in the slugs were calculated according to the calibration curve, and averaging was performed over 15 concentration fields for the studied flow rates in each position. Afterward, with the help of a script written in Python, the average values of the concentrations in the slugs and their standard deviations were calculated. Since the measured concentration field in the slug is depth-averaged, we numerically averaged concentration fields point by point, obtaining the average volumetric concentration. An example of the evolution of concentration fields of solute in slugs of the continuous phase for the fixed dispersed phase flow rate Q_d = 3 μL/min and different flow rates of the continuous phase downstream from the flow is shown in Figure 7.

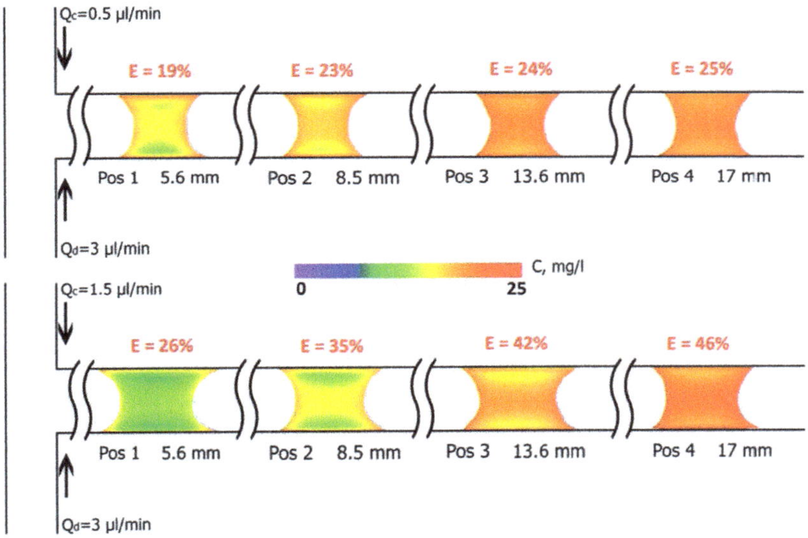

Figure 7. The evolution of concentration fields with increasing Q_c at a fixed Q_d = 3 μL/min.

4. Discussion

As a result of the measurement of the concentration fields in the slugs of the continuous phase, the dependences of the extraction efficiency and the overall volumetric mass transfer coefficient on the residence time and the total flow rate of the phases were obtained. The calculation was made using Equations (1) and (2) and the equilibrium solute concentration in the continuous phase for the studied flow rates was determined by Equation (6). The residence time τ was assessed as the distance from the T-junction to the measurement area L divided by plug velocity U_{plug}.

Figure 8 shows the dependence of the extraction efficiency and the mass transfer coefficient at a fixed flow rate ratio $Q_d/Q_c = 2$. The error bars correspond to the standard deviation of the values calculated based on 15 slugs. The extraction efficiency and the overall volumetric mass transfer coefficient were found to increase with an increase in the total flow rate. We assume that is due to the increasing intensity of vortices inside the slugs and plugs, providing a higher concentration gradient at the interfacial area between phases.

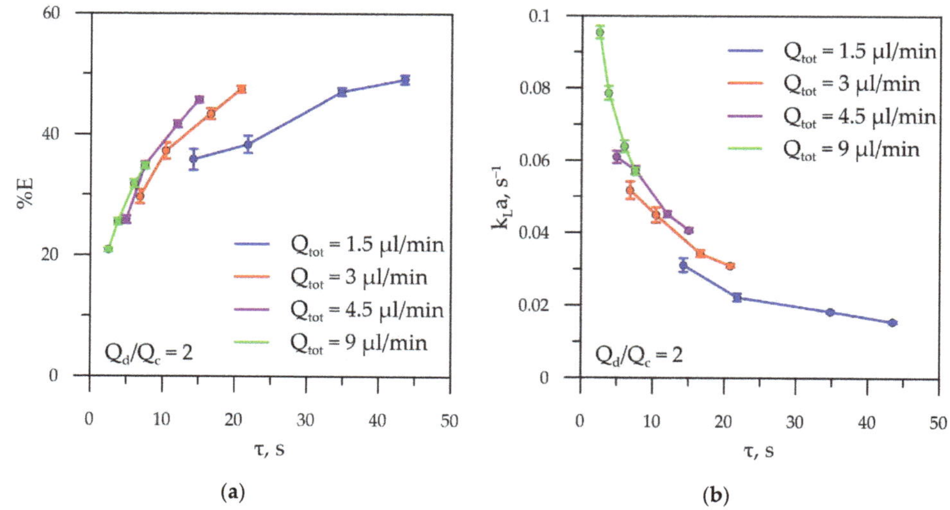

Figure 8. (**a**) Extraction efficiency and (**b**) overall volumetric mass transfer coefficient as a function of residence time at a fixed flow rate ratio.

Figure 9 shows the dependence of the extraction efficiency at 17 mm downstream of the T-junction and the overall volumetric mass transfer coefficient at the distance of 5.6 mm from the T-junction on the total flow rate of the phases for various flow rate ratios. Increasing the flow rate ratio Q_d/Q_c at a fixed total flow rate Q_{tot} decreases the extraction efficiency and mass transfer coefficient, possibly due to the fact that increasing Q_d/Q_c leads to an increase in the equilibrium concentration but the average concentration in the slug remains approximately the same. For the range of studied flow rates, the highest mass transfer efficiency was achieved at a minimal studied flow rate ratio Q_d/Q_c close to unity.

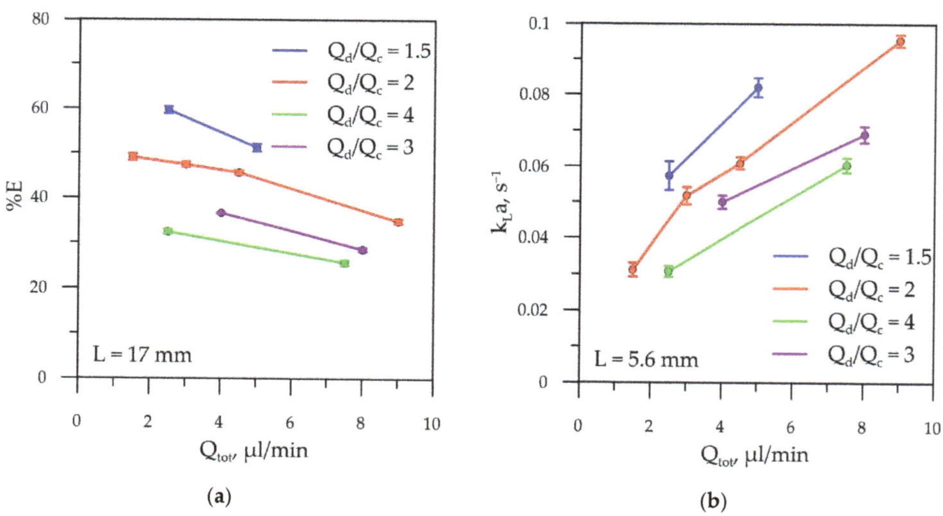

Figure 9. (**a**) Extraction efficiency at 17 mm from the T-junction as a function of the total phase flow rate at different phase ratios; (**b**) overall volumetric mass transfer coefficient at 5.6 mm from the T-junction as a function of the total phase flow at different phase ratios.

We correlated the obtained values of the total volume mass transfer coefficient with the model from Kashid et al. [13], which was developed to assess the mass transfer intensity between microchannel input and output. The authors used a fixed flow rate ratio and considered the part containing the length of the plug to be a constant. In our case, the flow rate ratio is changing, so we cannot neglect this part of the model:

$$k_L a \cdot \tau = a \cdot (Ca)^{b} \cdot (Re)^{c} \cdot \left(\frac{D_h}{L}\right)^{d} \cdot \left(\frac{L_{plug}}{D_h}\right)^{e} \tag{10}$$

where $Ca = \frac{\mu_M U_{plug}}{\sigma}$—capillary number and $Re = \frac{\rho_M U_{plug} D_h}{\mu_M}$—Reynolds number, which includes dynamic viscosity (μ_M) and density (ρ_M) of the mixture, D_h—hydraulic diameter of the microchannel, L—length between the beginning of the channel and a point downstream of the T-junction, and L_{plug}—plug length. Mixture properties were calculated by the following equations:

$$\mu_M = \left(\frac{\varepsilon_c}{\mu_c} + \frac{\varepsilon_d}{\mu_d}\right)^{-1} \tag{11}$$

$$\rho_M = \left(\frac{\varepsilon_c}{\rho_c} + \frac{\varepsilon_d}{\rho_d}\right)^{-1}, \tag{12}$$

where ε_c and ε_d—phase fraction of the continuous and dispersed phases.

Constants $a - e$ are adjustable parameters determined by fitting experimental data to the correlation (10) using the Levenberg–Marquardt algorithm. As a result, the following parameter values were obtained:

$$k_L a \cdot \tau = 0.042 \cdot (Ca)^{-0.92} \cdot (Re)^{0.64} \cdot \left(\frac{D_h}{L}\right)^{-0.59} \cdot \left(\frac{L_{plug}}{D_h}\right)^{-0.04} \tag{13}$$

The experimental data versus the model prediction are plotted in Figure 10. The R-squared value was 0.966. The median deviation between the experimental and predicted data was 8%. The analysis of the obtained parameters shows the dependence on the residence time and the total flow rate to be aligned with the experimentally obtained results and previously conducted studies [20,34]. The obtained coefficients correspond to the dependencies on the flow-rate ratio and total flow rate in Bai et al. [11]. At the same time, compared to the values of the parameters in Kashid et al. [7], the discrepancy in the dependence from the Reynolds number is observed.

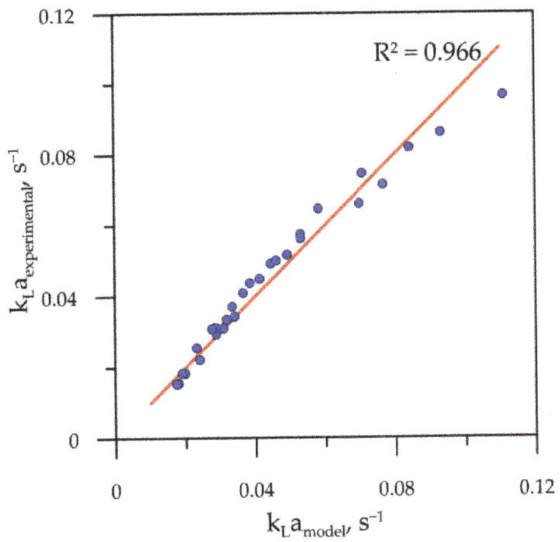

Figure 10. Experimental data of overall volumetric mass-transfer coefficient versus the data predicted by the model.

5. Conclusions

Experiments were conducted to investigate mass transfer during the flow of immiscible liquids in a T-type microchannel using the micro-LIF method. The concentration fields of the fluorescent dye in the slugs of the continuous phase were measured. To calculate the extraction efficiency and the overall volumetric mass transfer coefficient, the partition coefficient for the studied set of liquids was experimentally determined.

The results of the experiments led to conclusions regarding the effect of the total flow rate and the phase flow ratio on the efficiency of mass transfer and overall volumetric mass transfer coefficient. It was found that an increase in the flow rate ratio leads to a decrease in mass transfer efficiency. Meanwhile, an increase in the total flow rate results in an increase in both the extraction efficiency and mass transfer coefficient. A model for the overall volumetric mass transfer coefficient was also developed with an R^2 value of 0.966. This model supports the conclusions about the impact of the total flow rate and flow-rate ratio on the intensity of mass transfer between phases.

Author Contributions: Conceptualization, A.Y.; methodology, A.Y. and A.K.; investigation, S.V.; resources, A.Y.; data curation, S.V.; writing—original draft preparation, S.V. and A.Y.; writing—review and editing, A.B. and A.K.; supervision, A.B.; project administration, A.Y.; funding acquisition, A.Y. and A.B. All authors have read and agreed to the published version of the manuscript.

Funding: The slug flow properties study (Section 3.1) was supported by a grant from the Russian Science Foundation (project No 21-79-10307). Mass transfer measurements (Section 3.2) have been financially supported by the Ministry of Science and Higher Education of the Russian Federation, Project No 075-15-2022-1043.

Data Availability Statement: The data that support the findings of this study are available from the corresponding author upon reasonable request.

Conflicts of Interest: The authors declare no conflict of interest.

References

1. Abdollahi, A.; Sharma, R.N.; Vatani, A. Fluid flow and heat transfer of liquid-liquid two phase flow in microchannels: A review. *Int. Commun. Heat Mass Transf.* **2017**, *84*, 66–74. [CrossRef]
2. Kandlikar, S.G.; Garimella, S.; Li, D.; Colin, S.; King, M.R. *Heat Transfer and Fluid Flow in Minichannels and Microchannels*; Elsevier: Amsterdam, The Netherlands, 2014; ISBN 9780080983462.
3. Chen, G.-B.; Chao, Y.-C.; Chen, C.-P. Enhancement of hydrogen reaction in a micro-channel by catalyst segmentation. *Int. J. Hydrogen Energy* **2008**, *33*, 2586–2595. [CrossRef]
4. Wang, K.; Li, L.; Xie, P.; Luo, G. Liquid–liquid microflow reaction engineering. *React. Chem. Eng.* **2017**, *2*, 611–627. [CrossRef]
5. Priest, C.; Zhou, J.; Klink, S.; Sedev, R.; Ralston, J. Microfluidic Solvent Extraction of Metal Ions and Complexes from Leach Solutions Containing Nanoparticles. *Chem. Eng. Technol.* **2012**, *35*, 1312–1319. [CrossRef]
6. Tran, T.M.; Lan, F.; Thompson, C.S.; Abate, A.R. From tubes to drops: Droplet-based microfluidics for ultrahigh-throughput biology. *J. Phys. D Appl. Phys.* **2013**, *46*, 114004. [CrossRef]
7. Chen, R.; Su, G.H.; Zhang, K. Analysis on the high-quality development of nuclear energy under the goal of peaking carbon emissions and achieving carbon neutrality. *Carbon Neutrality* **2022**, *1*, 1–12. [CrossRef]
8. Voumik, L.C.; Science, N. Impact of Renewable and Non-Renewable Energy on EKC in SAARC Countries: Augmented Mean Group Approach. *Energies* **2023**, *16*, 2789. [CrossRef]
9. Swanson, J.L. PUREX Process Flowsheets. In *Science and Technology of Tributyl Phosphate*; Schulz, W.W., Burger, L.L., Navratil, J.D., Bender, K.P., Eds.; CRC Press: Boca Raton, FL, USA, 1984; p. 55.
10. Ma, S.; Sherwood, J.M.; Huck, W.T.S.; Balabani, S. On the flow topology inside droplets moving in rectangular microchannels. *Lab Chip* **2014**, *14*, 3611–3620. [CrossRef]
11. Abiev, R.S.; Butler, C.; Cid, E.; Lalanne, B.; Billet, A.-M. Mass transfer characteristics and concentration field evolution for gas-liquid Taylor flow in milli channels. *Chem. Eng. Sci.* **2019**, *207*, 1331–1340. [CrossRef]
12. Butler, C.; Cid, E.; Billet, A.-M. Modelling of mass transfer in Taylor flow: Investigation with the PLIF-I technique. *Chem. Eng. Res. Des.* **2016**, *115*, 292–302. [CrossRef]
13. Kashid, M.N.; Gupta, A.; Renken, A.; Kiwi-Minsker, L. Numbering-up and mass transfer studies of liquid–liquid two-phase microstructured reactors. *Chem. Eng. J.* **2010**, *158*, 233–240. [CrossRef]
14. Dietrich, N.; Loubière, K.; Jimenez, M.; Hébrard, G.; Gourdon, C. A new direct technique for visualizing and measuring gas–liquid mass transfer around bubbles moving in a straight millimetric square channel. *Chem. Eng. Sci.* **2013**, *100*, 172–182. [CrossRef]

15. van Baten, J.M.; Krishna, R. Corrigendum to "CFD simulations of mass transfer from Taylor bubbles rising in circular capillaries". *Chem. Eng. Sci.* **2004**, *59*, 2535–2545. [CrossRef]
16. Yue, J.; Luo, L.; Gonthier, Y.; Chen, G.; Yuan, Q. An experimental study of air–water Taylor flow and mass transfer inside square microchannels. *Chem. Eng. Sci.* **2009**, *64*, 3697–3708. [CrossRef]
17. Kashid, M.; Renken, A.; Kiwi-Minsker, L. Influence of Flow Regime on Mass Transfer in Different Types of Microchannels. *Ind. Eng. Chem. Res.* **2011**, *50*, 6906–6914. [CrossRef]
18. Zhang, Q.; Liu, H.; Zhao, S.; Yao, C.; Chen, G. Hydrodynamics and mass transfer characteristics of liquid–liquid slug flow in microchannels: The effects of temperature, fluid properties and channel size. *Chem. Eng. J.* **2019**, *358*, 794–805. [CrossRef]
19. Yao, C.; Ma, H.; Zhao, Q.; Liu, Y.; Zhao, Y.; Chen, G. Mass transfer in liquid-liquid Taylor flow in a microchannel: Local concentration distribution, mass transfer regime and the effect of fluid viscosity. *Chem. Eng. Sci.* **2020**, *223*, 115734. [CrossRef]
20. Angeli, P.; Tsaoulidis, D.; Weheliye, W.H. Studies on mass transfer of europium(III) in micro-channels using a micro Laser Induced Fluorescence technique. *Chem. Eng. J.* **2019**, *372*, 1154–1163. [CrossRef]
21. Ganguli, A.A.; Pandit, A.B. Hydrodynamics of Liquid-Liquid Flows in Micro Channels and Its Influence on Transport Properties: A Review. *Energies* **2021**, *14*, 6066. [CrossRef]
22. Bai, L.; Zhao, S.; Fu, Y.; Cheng, Y. Experimental study of mass transfer in water/ionic liquid microdroplet systems using micro-LIF technique. *Chem. Eng. J.* **2016**, *298*, 281–290. [CrossRef]
23. Kuhn, S.; Jensen, K.F. A pH-Sensitive Laser-Induced Fluorescence Technique to Monitor Mass Transfer in Multiphase Flows in Microfluidic Devices. *Ind. Eng. Chem. Res.* **2012**, *51*, 8999–9006. [CrossRef]
24. Cheng, N.-S. Formula for the Viscosity of a Glycerol−Water Mixture. *Ind. Eng. Chem. Res.* **2008**, *47*, 3285–3288. [CrossRef]
25. Volk, A.; Kähler, C.J. Density model for aqueous glycerol solutions. *Exp. Fluids* **2018**, *59*, 75. [CrossRef]
26. Demond, A.H.; Lindner, A.S. Estimation of interfacial tension between organic liquids and water. *Environ. Sci. Technol.* **1993**, *27*, 2318–2331. [CrossRef]
27. Rader, C.A.; Schwartz, A.M. The Migration of Liquids in Textile Assemblies: Two-Component and Two-Phase Liquid Systems. *Text. Res. J.* **1962**, *32*, 140–153. [CrossRef]
28. Yagodnitsyna, A.A.; Kovalev, A.V.; Bilsky, A.V. Flow patterns of immiscible liquid-liquid flow in a rectangular microchannel with T-junction. *Chem. Eng. J.* **2016**, *303*, 547–554. [CrossRef]
29. Kovalev, A.V.; Yagodnitsyna, A.A.; Bilsky, A.V. Viscosity Ratio Influence on Liquid-Liquid Flow in a T-shaped Microchannel. *Chem. Eng. Technol.* **2020**, *44*, 365–370. [CrossRef]
30. Kovalev, A. The Influence of Viscosity on the Hydrodynamics of Immiscible Liquid-Liquid Flows in Rectangular Microchannels. Ph.D. Thesis, Kutateladze Institute of Thermophysics SB RAS, Novosibirsk, Russia, 2022.
31. Garstecki, P.; Fuerstman, M.J.; Stone, H.A.; Whitesides, G.M. Formation of droplets and bubbles in a microfluidic T-junction—Scaling and mechanism of break-up. *Lab Chip* **2006**, *6*, 437–446. [CrossRef]
32. Xu, J.H.; Li, S.W.; Tan, J.; Luo, G.S. Correlations of droplet formation in T-junction microfluidic devices: From squeezing to dripping. *Microfluid. Nanofluidics* **2008**, *5*, 711–717. [CrossRef]
33. Zehentbauer, F.M.; Moretto, C.; Stephen, R.; Thevar, T.; Gilchrist, J.R.; Pokrajac, D.; Richard, K.L.; Kiefer, J. Fluorescence spectroscopy of Rhodamine 6G: Concentration and solvent effects. *Spectrochim. Acta Part A Mol. Biomol. Spectrosc.* **2014**, *121*, 147–151. [CrossRef]
34. Qian, J.-Y.; Li, X.-J.; Wu, Z.; Jin, Z.-J.; Sunden, B. A comprehensive review on liquid–liquid two-phase flow in microchannel: Flow pattern and mass transfer. *Microfluid. Nanofluidics* **2019**, *23*, 1–30. [CrossRef]

energies

Article

Combining Microstructured Surface and Mesh Covering for Heat Transfer Enhancement in Falling Films of Refrigerant Mixture

Oleg Volodin *, Nikolay Pecherkin and Aleksandr Pavlenko *

Kutateladze Institute of Thermophysics, Siberian Branch of the Russian Academy of Sciences, 1 Lavrentyev Ave., 630090 Novosibirsk, Russia
* Correspondence: volodin@ngs.ru (O.V.); pavl@itp.nsc.ru (A.P.)

Abstract: The article presents the experimental results of combining a basic microstructure with partly closed pores and a mesh covering for heat transfer enhancement at the film flow of a refrigerant mixture. To reveal the effect of the combined structure, heat transfer on a microstructured surface without a covering as well as on a smooth surface with a mesh covering only has been studied. All experimental series were carried out using a binary mixture of R114 and R21 refrigerants. The mixture film flowed down the outer surface of a vertical cylinder in the undeveloped turbulence regime, when the film Reynolds number varied from 400 to 1300. It is shown that a microstructured surface with a fin pitch of 200 μm, fin height of 220 μm, and longitudinal knurling pitch of 160 μm, created by deformational cutting, demonstrates significant heat transfer enhancement: up to four times as compared to a smooth surface. However, adding a mesh covering with an aperture of 220 μm and a wire diameter of 100 μm reduces the intensification. The mesh covering overlaid on a smooth surface also does not provide heat transfer enhancement as compared to the smooth surface itself. The absence or even deterioration of heat transfer enhancement on surfaces with mesh covering can be primarily associated with the low thermal conductivity of the mesh material and shortcomings of the applied method of mesh mounting. The possibility of deteriorating vapor removal due to the incorrect selection of mesh covering parameters was also analyzed. The heat transfer coefficient values obtained for basic microstructured surfaces were compared with the dependencies available in the literature for predicting pool boiling heat transfer on microfinned surfaces.

Keywords: nucleate boiling; combined coating; deformational cutting; mesh covering; falling films; refrigerant mixture

Citation: Volodin, O.; Pecherkin, N.; Pavlenko, A. Combining Microstructured Surface and Mesh Covering for Heat Transfer Enhancement in Falling Films of Refrigerant Mixture. *Energies* **2023**, *16*, 782. https://doi.org/10.3390/en16020782

Academic Editors: Fedor Ronshin and Vladimir Serdyukov

Received: 29 November 2022
Revised: 5 January 2023
Accepted: 6 January 2023
Published: 10 January 2023

1. Introduction

The study of heat transfer in falling films of liquids and their mixtures is important for improving the efficiency of numerous systems using film flows—from natural gas liquefaction (LNG) plants, distillation plants, and absorption apparatuses, as well as evaporative equipment of the chemical and food industries—to desalination plants and electronic equipment cooling systems (e.g., spray and falling-film cooling system [1]). The intensification of heat transfer at boiling and evaporation of liquids by structuring the heat-generating surface is a key method for increasing the efficiency of both traditional [2–5] and renewable [6] energy systems. Despite the fact that studies on heat transfer enhancement have been actively carried out since the middle of the last century, their relevance at the current pace of technology development is only increasing. Among various types of structured coatings (including capillary-porous), wire mesh coverings—which can be attributed to the simplest porous coatings—are among the most accessible and easily modified. The advantages of these coverings also include: ease of installation, high scalability, low production cost, and reproducibility of the geometric parameters of wire meshes. All this determines the renewed interest in recent years in the use of mesh coverings in various configurations [5].

For a rather long period of research on the use of mesh coverings of heating surfaces for heat transfer enhancement, i.e., from about 1975 and until the end of the first decade of the 21st century, not so many promising results were obtained on increasing the heat transfer coefficient (HTC) or delaying of reaching the critical heat flux (CHF) [5]. It was generally believed that mesh coverings enhance heat transfer at low heat fluxes and impair it at high heat fluxes. The enhancement was associated with the increased area of the mesh-covered surface and consequently the higher number of active nucleation sites, whereas at higher heat fluxes heat transfer becomes dependent on the vapor removal, and in this case, the mesh coatings (especially multilayer ones) impair heat transfer [5,7].

However, the results of studies conducted over the past decade [5] show that mesh coverings can be an effective means of heat transfer enhancement in a wide range of heat flux changes. Some authors have also expanded previous ideas [7] about the causes of the HTC improvement by introducing fresh concepts like "micro-chimney effects" and "gradient mesh coatings" [6]. The desired enhancement can be realized provided that the mesh geometrical parameters, the number of mesh layers, the wire material, and the mesh mounting method are optimally chosen. Thus, significant heat transfer enhancement was obtained using the mentioned above gradient mesh coatings [6] (HTC enhancement up to 6.6 times), as well as uniform layers mesh coatings at atmospheric (HTC enhancement from about 3 [8] to 10 times [9]) and subatmospheric pressures (HTC enhancement up to 22 times [9]). Promising results have also been achieved by combining mesh covering with other types of surface treatment: microfinning, nanostructuring, mesh with cells filled with powder from micro-nanoparticles, as well as when processing wire mesh itself.

Below we briefly review the works devoted to the study of heat transfer enhancement and the increase of the critical heat flux using multilayer or combined mesh coatings, which contain the most interesting or promising results in the opinion of the authors.

In the works [10,11], it was demonstrated that covering a rough surface with a mesh can reduce the intensity of heat transfer. Trying to use simple and practical methods of enhancement, Tsay et al. [10] were among the first to combine the roughening of heat transfer surface with a mesh covering (from stainless steel AISI 304) to enhance boiling in thin layers of water on a horizontal plate (also from AISI 304) of 10 cm length, 2.5 cm width, and 0.1 cm thickness. The authors also investigated the effect of applying various mesh coverings (mesh 16, 24, and 50) to a smooth plate. As a result, it received up to a sevenfold HTC increase (at liquid level H = 5 mm and mesh 16) at low heat fluxes. The thinner the liquid layer, the greater the observed enhancement. However, no additional enhancement for the mesh-covered rough surfaces as compared with the mesh-covered smooth ones was received.

The use of combined coatings by Brautsch and Kew [11] was also unsuccessful in terms of heat transfer enhancement. The authors used low carbon stainless steel AISI 304L meshes (mesh 50, 100, 150, and 200), as well as two heat-releasing surfaces with R_a = 0.42 μm and R_a = 1.04 μm, to study heat transfer enhancement in the saturated pool boiling of water on the vertical test section. They showed that both techniques—roughening of surface and covering the heater with a mesh, are effective. However, combining them was shown to be ineffective, resulting in the HTC deterioration even as compared to the smooth surface.

Despite these and other unsuccessful experiments on heat transfer enhancement [5], changes in the material and method of mounting mesh coverings allowed Sloan et al. [9] to achieve significant intensification of HTC over the entire range of heat fluxes. The authors [9] studied the pool boiling of water at subatmospheric pressure on a 4 cm^2 vertically oriented copper circular disk, covered with 1–8 layers of copper mesh. It was shown that eight-layer mesh 145 covering demonstrated about a 10-fold HTC increase (at ΔT = 10 K) at atmospheric pressure and a 22-fold increase (at ΔT = 8 K) at subatmospheric pressure (0.2 atm). It is worth noting that Sloan et al. used copper mesh coverings previously cold-rolled to increase the available surface area for the diffusion bonding process by producing flat spots on the high points of each wire.

Chien and Tsai [12] studied heat transfer at film flow and pool boiling of R-245fa at different saturation temperatures on horizontal finned copper tubes (of 0.4 mm fin height, 60 FPI) and the same tubes, covered with copper mesh. They achieved a notable HTC enhancement—up to four and seven times at pool boiling of R-245fa for 5 and 20 °C, respectively, and up to five times at film flow of R-245fa, compared to a smooth tube. For comparison, the uncovered finned tube in the latter case provided intensification only 3.5 times.

The authors [13] used a combined surface with brass or copper meshes (mesh 80, 100, and 120) covered finned horizontal tubes with fins of 0.2–0.4 mm high. Pool boiling of R-134a at saturation temperatures of 5, 10, and 26.7 °C was enhanced 2–3 times. The best performance was achieved by wrapping a brass mesh 100 on a tube with 0.4 mm fin height—up to 8 times (peak value) compared to an uncovered tube.

Kim et al. [14] achieved 84% CHF enhancement by using single-layered stainless steel mesh (with wire diameter 0.29 ÷ 0.7 mm and mesh aperture 1.29 ÷ 2.67 mm range) with micro/nano-sized pore structure of meshes applied to SiO_2 heating surface at water pool boiling. The authors underlined that their method of increasing CHF does not require any modifications of the heating surface and can be easily adopted in different technical applications (for example, to create IC chip coatings).

Dąbek et al. [15] studied the pool boiling of water and ethyl alcohol on a horizontal copper heater of 30 mm diameter with copper or bronze one- or two-layer mesh coatings at atmospheric pressure. The possibility of sevenfold and fourfold enhancement as compared to a smooth surface (at superheat of about 8 K) for copper meshes with apertures of 0.75 mm and 0.2 mm, respectively, has been demonstrated.

The work of [16] is notable for the fact that the authors used 3D-printed mesh structures for heat transfer enhancement at the pool boiling of water at the saturation line. A total of 12 samples were divided into two groups of printed meshes: "thin" (0.75 mm wall height) and "thick" (1.5 mm wall height), with pitch varied in a range of 0.4 ÷ 1.3 mm. Stainless steel 316 L powder was used in the process of selective laser melting (SLM). Zhang et al. [16] showed the possibility of a threefold CHF enhancement as compared to a smooth surface by applying a mesh with 1.1 mm pitch ("thick"), also HTC enhancement of two to three times by applying meshes with 0.5 mm pitch ("thin") and 0.7–1.1 mm pitch ("thick") as compared to the smooth surface was achieved. This work [16] demonstrates the perspectives of additive manufacturing (AM) for creating prototypes of samples with precisely controlled structure parameters.

Pastuszko et al. [17] used micro-finned surfaces with copper mesh covering as well as micro-finned surfaces covered by copper perforated foil at pool boiling of water, ethanol, Novec-649, and FC-72 at atmospheric pressure. Microfins covered with wire mesh produced the highest HTC among studied surfaces at medium and high heat fluxes for water, low and medium heat fluxes for ethanol, and medium heat fluxes for FC-72.

The authors of [6] have demonstrated that multilayer mesh coatings can be highly efficient intensifiers by studying the pool boiling of distilled water at the saturation line on a heated surface covered with multi-layer mesh. Four configurations of six-layer copper mesh coverings with gradient (direct or inverse) or uniform porosity were studied. It was shown that a six-layer mesh (3 + 3) with coarser three upper layers gives a maximum HTC enhancement—up to 6.6 times (261 kW/m^2K), along with a three-fold enhancement in CHF (outstanding 2719 kW/m^2). The authors associate the obtained enhancement results with so-called "micro-chimney effects" [6], taking place in gradient porous micro meshes.

Huang et al. [18] applied four hybrid surfaces for heat transfer enhancement at subcooled water flow boiling in channels (at a pressure of 0.5 MPa, a flow velocity of 1–5 m/s, and an inlet temperature of 298 K). The authors used wire mesh coatings combined with a powder mixture of micro/nanoparticles of Ag, Cu, and Ti. Heat flux removed by combined surfaces was two to three times higher than heat flux removed by the smooth surface, CHF for the investigated surfaces increased by 80–200%.

The authors [19], as well as [6] demonstrate the effectiveness of multilayer gradient meshes, but in terms of enhancement of wicking capability. It was shown that the wicking capability of a multilayer gradient mesh consisting of three lower layers of a mesh 100 and three upper layers of a mesh 300 is significantly enhanced compared to wicks consisting of a multilayer mesh with identical layer characteristics.

In the previously mentioned work [8] the authors proposed a surface, sintered with multilayer copper meshes having identical geometrical characteristics (mesh 200 with 30 μm wire diameter), studying the water pool boiling. It is shown that an increase in the number of layers (up to 5) can reduce the size of the micropores, increasing the density of nucleation sites and improving capillary wicking performance, thus improving the HTC and delaying the boiling crisis development. The multilayer mesh with 5 layers demonstrates optimal boiling performance, providing the highest CHF of 208 W/cm^2 and the highest HTC of 16 W/(cm^{2}K) The authors made the conclusion that the remarkable boiling performance along with the low cost, simplicity, and high durability of mesh coatings show the industrial prospects for commercial compact microelectronics cooling.

Hu et al. [20] along with the authors of [8,9] demonstrate the perspectives of using uniform multilayer copper micromeshes for heat transfer enhancement, in particular, to increase heat transfer proportion of liquid film boiling in spray cooling of 10 × 10 mm^2 target surface. It is shown that a four-layer mesh 100 covering (mesh covering with 50 μm wire diameter and 204 μm aperture) fabricated by diffusion bonding, exhibits the best heat transfer performance with CHF of 605 W/cm^2 and maximum HTC of 71 kW/(m^{2}K), which corresponds to enhancing by 127% and 176%, compared with the uncovered surface, respectively.

Thus, the above shows that wire mesh coatings in various combinations and modifications can be an effective means of boiling heat transfer for a wide spectrum of technical applications, including apparatuses using pool boiling [6–9,11–17], working at low pressures [9], using spray cooling [20], thin layers of liquid [10], film flow [7,12], or microchannels [18]. The aim of this work is to initiate the investigation of the efficiency of combined coatings for heat transfer enhancement in the binary refrigerant mixture films falling down the outer surface of a vertical cylinder. Despite the fact that the cooling of heating surfaces by falling films does not allow the removal of large heat fluxes as in the case of the recognized leader among cooling methods—spray irrigation, which makes it possible to remove heat fluxes up to 1000 W/cm^2 [21], tubular heat exchangers operating at low and moderate heat fluxes are widely used and in demand in the industry (for example, in LNG systems), and the possibility of enhancing heat transfer and improving the ergonomics of film evaporators by using tubes with modified surfaces requires systematic research in this field.

2. Experimental Setup and Procedure

2.1. Test Sections

To create the basic microstructure with halfway-closed micropores, providing more effective nucleate boiling, the previously well-established method of deformational cutting (MDC) was used [22–24]. The microstructure, created on the outer surface of a copper cylinder with a diameter of 50 mm and wall thickness of 1.5 mm, has the next geometrical parameters: fin pitch—200 μm, fin height—220 μm, and longitudinal knurling pitch—160 μm (Figure 1a,b). The knurling over microfinning, crushing the tops of the fins with a decrease in its height by 35%, is carried out in order to create the halfway-closed pores surface. In the work of [22], it was shown the effectiveness of such structure type for nucleate boiling enhancement. In this study, we slightly modified the parameters of the most effective MDC-surface (microstructure No. 1 from [22]), increasing its fin pitch (from 100 μm to 200 μm)—with the aim of improving vapor removal, and reducing its knurling pitch (from 318 μm to about 160 μm)—in attempt to increase the amount of nucleation centers.

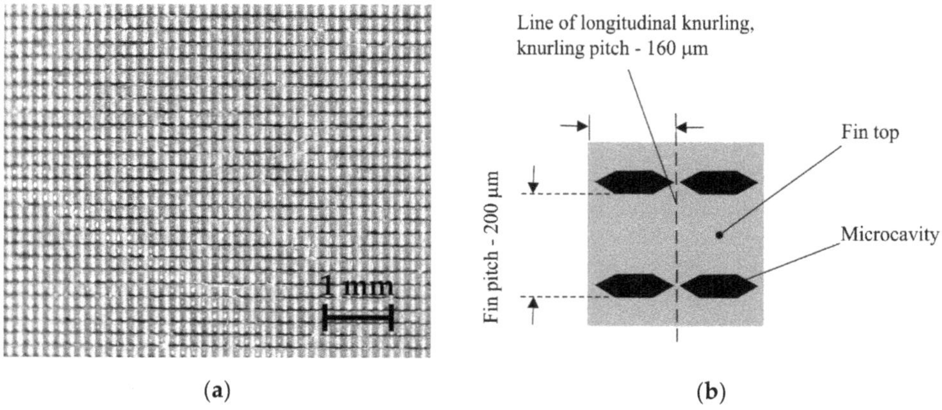

Figure 1. (**a**) photograph of the surface, structured by deformational cutting, (**b**) sketch of the element of the MDC-surface.

At the next stage of test section preparation, the stainless steel mesh (AISI 304) with an aperture of 220 μm and wire diameter of 100 μm was tightly wrapped around the microstructured tube, fixed, and then soldered with a slight overlap of layer upon layer. Resulting in a combined structure (Figure 2a), the single element of which is shown (according to scale) in Figure 2b. The characteristics of mesh were chosen so as to approximately correspond to the parameters of the basic structured surface since a finer upper mesh would impair vapor removal, while a too-coarse upper mesh would not have an effect on the nucleate boiling process.

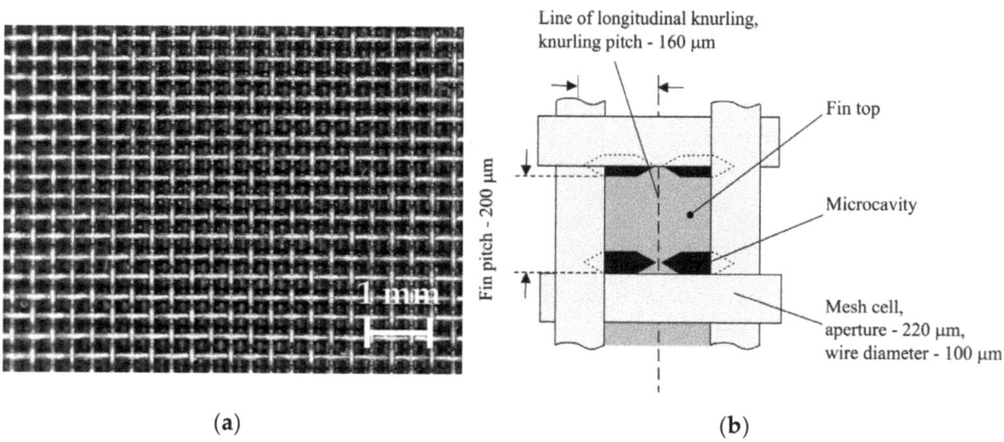

Figure 2. (**a**) photograph of the combined surface, (**b**) sketch of the element of the combined surface.

In addition to applied steel mesh coverings (which did well in [10] and some other works, despite the low thermal conductivity of steel), in the next stage of the study, it is planned to use copper and brass meshes; the geometric parameters of meshes and applied installation methods will also vary.

To reveal the effect of the combined coating structure on heat transfer intensification, the heat transfer on the microstructured surface without mesh covering as well as on a smooth surface with mesh covering alone and on an uncovered smooth surface were also studied. All types of test sections used in the study are listed in Table 1.

Table 1. Types of test sections used in the study.

Test Section Type	Material	Geometrical Parameters
Smooth tube	Copper M1	smooth tube with roughness R_a = 2.5 μm
Microstructured tube	Copper M1	microstructure with fin pitch 200 μm, fin height 220 μm and knurling pitch 160 μm
Smooth tube + mesh	Copper M1 + AISI 304	smooth tube with R_a = 2.5 μm + mesh with aperture 220 μm and wire diameter 100 μm
Microstructured tube + mesh	Copper M1 + AISI 304	microstructure with fin pitch 200 μm, fin height 220 μm and knurling pitch 160 μm + mesh with aperture 220 μm and wire diameter 100 μm

2.2. Experimental Setup and Parameters

The experiments were carried out in the test column (1) 1.5 m high and with an inner diameter of 0.27 m under saturation conditions at an absolute pressure of about 2 bar (Figure 3a,b). A more complete arrangement of the experimental setup is given in [25]. The R114/R21 refrigerant mixture is fed into a constant-level tank (2) from where it comes through the slot distributor to the test section (3). The local temperatures of the heating wall are controlled by five copper-constantan thermocouples (4) with a wire diameter of 0.15 mm, embedded flush with the wall surface and located vertically along the heated area having a length of 70 mm with a step of 13 mm. Thermocouples are handcrafted and calibrated over a range of operating temperatures (20–55 °C) using the proven reference thermometer (LT-300). The diameter of the thermocouple bead does not exceed 0.7 mm. The sensitivity of copper-constantan (CuKn) thermocouples or type T thermocouples is about 40 microvolts per degree. The use of a highly sensitive voltmeter APPA 207 made it possible to measure thermo-EMF values with an accuracy of ±8 μV. The cold junctions of thermocouples (5) are located at the bottom of the column, immersed in the liquid layer. The temperatures in the different sections of the column are controlled by platinum thermoresistors HEL-711-U-0-12-00 (6) by Honeywell. The HEL-711 temperature sensors allow the measurement of temperature in the range of −200 to +260 °C with an accuracy of ±0.1%. One of the thermistors located at the bottom of the column together with cold junctions of thermocouples in order to measure the temperature of the liquid to correctly build the dependence of the wall temperature on the thermo-EMF, taking into account the temperature fluctuations of the liquid at the column bottom.

The observations and high-speed shooting (by Phantom VEO 410 L) of processes, developing in the heated area, are carried out through three quartz optical windows (7). The pressure in the test volume is measured using a manometer Metran-100 (8).

Stabilized supply source Mastech HY10010E was used to heat the test section, capable of providing heat fluxes up to critical. The load was raised stepwise, the readings were taken while moving up the heat flux branch after the heater temperature reached a stationary value (which took about 10–15 min, depending on the heat transfer regime). The binary mixture film flowed down the vertical cylinder in the undeveloped turbulence regime with the film Reynolds number at the inlet of the test section varied from 400 to 1300. The film Reynolds number was determined as:

$$Re = \frac{4G}{\pi d\mu}, \tag{1}$$

here G is mass rate, d—tube diameter, and μ—dynamic viscosity. The mass flow rate was measured by the Coriolis flow meter CORI-FLOW M55.

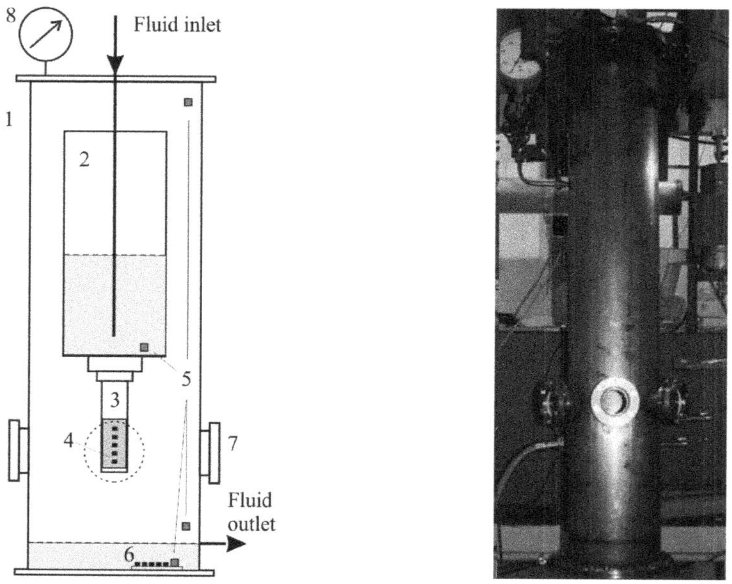

(**a**) (**b**)

Figure 3. The scheme of the test column (**a**): 1—sealed column; 2—constant level tank; 3—test section; 4—heated zone of the test section with thermocouples; 5—thermal resistors; 6—thermocouples cold junctions; 7—optical windows; 8—a manometer. The photograph of the test column without heat shield (**b**).

The distance from the slot distributor to the beginning of the heated area was equal to 100 mm, which noticeably exceeded the distance of hydrodynamic stabilization of the flow in all studied ranges of the flow rates; a corresponding estimation was carried out in [22].

The selected R114/R21 refrigerant mixture has low surface tension, high wettability, and low viscosity, which makes it convenient for modeling heat transfer processes for a wide class of low-viscosity technical fluids. In addition, the presence of the volatile component of the mixture (R114) can notably enhance the heat transfer process. In the work of [26], it was shown that for a 10% mixture of R114/R21, the heat flux density corresponding to the onset of boiling (ONB) was reduced by half as compared to pure R21, and an increase in HTC up to 2.5 times was observed in the range of heat fluxes q: $0.1 \div 10$ W/cm^2. It can also be noted that the use of a non-azeotropic mixture can increase the critical heat flux value, due to the later evaporation of the low-boiling component [27,28]. With this in mind, a mixture of R114 and R21 refrigerants with an initial concentration of volatile component R114 of about 12% (the mixture behaves as non-azeotropic at this concentration) was chosen as the working fluid.

2.3. Experimental Procedure and Data Reduction

Before the start of each experimental series (and after the tightness test), vacuum evacuation of the test volume was performed to degas the volume and the micropores of the test section. To evacuate the test setup (remove air) before filling it with the working fluid, an RV-40 forevacuum pump was used. Air was pumped out to a residual pressure of 6–8 Pa. After air was evacuated from the test setup, the setup volume was filled with vapors of the working fluid from the storage vessel. At room temperature of about (20–22) °C, the excess pressure in the setup is maintained at the level of (0.5–0.6) bar, which excludes air suction into the setup. Measurement of vapor pressure and vapor and liquid phase temperatures in the test volume made it possible to control the state of saturation of the working fluid and the absence of impurities (like air) during the experiments.

To measure the local wall temperatures, the thermo-EMF values from five thermocouples were recalculated into temperature values using a two-dimensional regression—the function of thermo-EMF and the temperature of cold junctions.

The heat flux density was calculated by the formula:

$$q = C\frac{I^2 R}{A}, \quad (2)$$

here, C is a correction factor taking into account heat losses due to the heat conductivity of the test section. According to the numerical calculations performed for the studied test sections, the heat loss due to longitudinal heat conductivity of cylinder walls did not exceed 5%. The areas of the modified heat-releasing surfaces A during calculations were taken equal to the area of the covered smooth surface.

The local heat transfer coefficients were determined by the standard formula:

$$h = \frac{q}{(T_w - T_s)}, \quad (3)$$

here T_W—temperature of the wall, T_S—saturation temperature.

The uncertainty in determining the local heat transfer coefficient h consisted of uncertainties in determining the heat flux q and the wall superheat temperature $\Delta T = (T_w - T_s)$.

The final uncertainty of the heat transfer coefficient $h = f(q, \Delta T)$ was calculated by the formula:

$$U^2(h) = \sum_{i=1}^{N} \left(\frac{\partial f}{\partial x_i}\right)^2 U^2(x_i) \quad (4)$$

or

$$\left(\frac{U(h)}{h}\right)^2 = \left(\frac{U(q)}{q}\right)^2 + \left(\frac{U(\Delta T)}{\Delta T}\right)^2, \quad (5)$$

where $U^2(\Delta T) = U^2(T_w) + U^2(T_s)$.

According to the assessment, the maximum error when determining ΔT was introduced by the relative uncertainty of T_w, measured by CuKn thermocouples, that did not exceed 10% at $\Delta T > 2$ °C. So the uncertainty of $\Delta T = (T_w - T_s)$ will also not exceed 10% up to the integer. The relative uncertainty of the heat flux q taking into account heat losses did not exceed 6%. Thus the relative uncertainty of the local heat transfer coefficient h did not exceed 12% at $\Delta T > 2$ °C.

3. Results and Analysis

Figure 4a–d illustrates the boiling process of the R114/R21 mixture falling film on studied test sections at the initial stages of nucleate boiling. The test sections structured by deformational cutting, both uncovered (Figure 4b) and having mesh covering (Figure 4d), demonstrate a more effective nucleation process than sections without basic MDC-structuring (Figure 4a,c), with nucleation sites activated from the beginning of heating area—so it becomes visible—to its end.

Figure 5a–d demonstrate the nucleate boiling of binary mixture falling film on studied sections at developed boiling regime—with the presence of very large bubbles (see for example Figure 5c) characteristic of the mixtures and according to our observations does not occur in pure refrigerants. Again it can be noted that both test sections with substrate structured by deformational cutting (Figure 5b,d) demonstrate a more effective nucleation process with nucleation sites evenly distributed over the entire heat-releasing surface. Next will be shown that the heat transfer data confirm the visual observation.

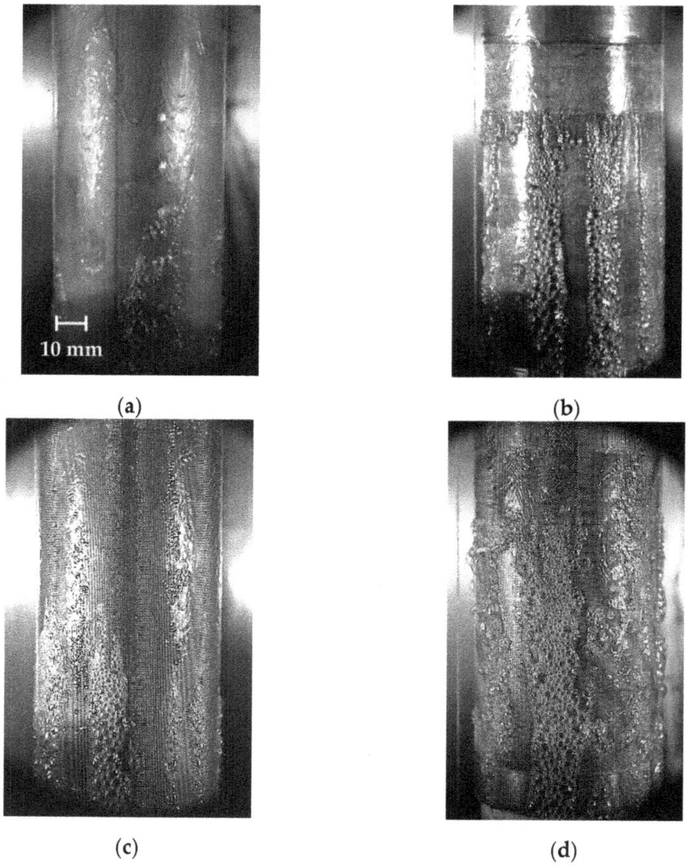

Figure 4. Inception of nucleate boiling of R114/R21 mixture on: (**a**) smooth (q = 1.7 W/cm^2, Re = 640), (**b**) microstructured by MDC (q = 1.7 W/cm^2, Re = 640), (**c**) smooth covered by mesh (q = 1.7 W/cm^2, Re = 845) and (**d**) combined (q = 1.7 W/cm^2, Re = 640) surfaces.

Boiling curves obtained for the investigated surfaces are shown in Figure 6a–d.

Figure 6a shows the boiling curve for the smooth surface. In the film evaporation regime ($q \leq 1$ W/cm^2), the flow rate does not affect heat transfer. For all Reynolds numbers, ONB takes place at $q \geq 1$ W/cm^2 and incipience superheat $\Delta T \approx 8$ K, respectively. Above $\Delta T \approx 8$ K and up to the developed boiling regime ($\Delta T \approx 10.5$ K), an insignificant effect of the flow rate on heat transfer is observed—the thinner the film, the higher heat transfer, as a rule. This could be due to the different contributions of evaporation to heat transfer, which takes place for different thicknesses of the falling film in the developing boiling regime. However, this effect is weakly expressed, and the presence of any kind of microtexture (Figure 6b–d) immediately shuffles the cards.

Microstructured surface by the method of DC (Figure 6b) demonstrates the greatest heat transfer efficiency, while the presence of the mesh covering does not affect (Figure 6c), or even worsens (Figure 6d), the heat transfer process. In the first case, we suppose, this is due to the fact that stainless steel (AISI 304) mesh used in the experimental series has a low thermal conductivity (15 W/mK), and the method of mesh attachment used is apparently not effective enough to create an ideal mesh-heated surface contact. In the case of combined surface, this also can be connected with the fact that we did not choose the mesh covering parameters well enough for our experimental conditions, which may cause the deterioration of vapor removal.

Thus, despite the fact that the presence of an applied steel mesh covering to some extent orders nucleate boiling (one can compare boiling patterns on mesh-covered surfaces in Figure 4c,d and on the smooth one in Figure 4a), taking into account the deterioration of heat transfer, at the next stage of the experiment, measures should take into account the existing shortcomings of the mesh coating used. It is planned to use highly thermally conductive copper mesh coatings and the process of sintering to ensure contact of the mesh covering with the heated surface; a more careful selection of mesh covering parameters also should be provided.

A comparison of experimental data on HTC and its enhancement on the studied surfaces for the range of Reynolds number 400 ÷ 1300 is presented in Figure 7a,b. As noted above, the presence of mesh with the given characteristics does not enhance the boiling process when covering the smooth surface: data on HTC coincide with data for a smooth surface, Figure 7a. The microstructured surface, created by deformational cutting (Figure 7b), demonstrates the greatest heat transfer enhancement (up to four times as compared with the smooth one). In the case of mesh overlay of the microstructured surface (i.e. in the case of combined coating), even impairment (up to two times) in heat transfer as compared to high-performance MDC-surface is observed, Figure 7b.

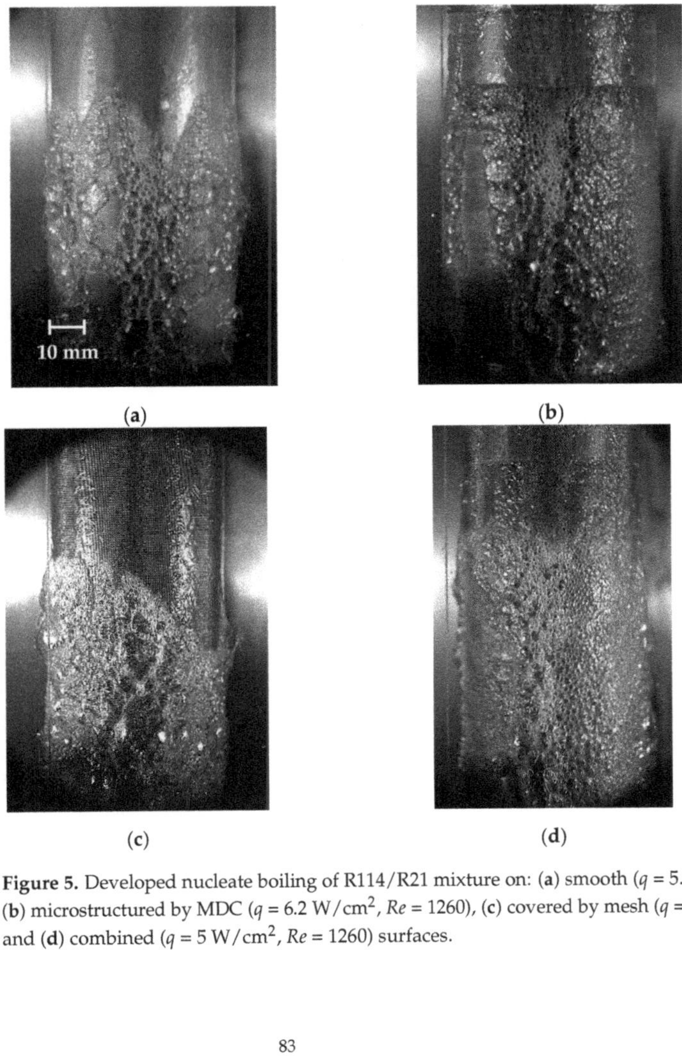

Figure 5. Developed nucleate boiling of R114/R21 mixture on: (**a**) smooth (q = 5.6 W/cm^2, Re = 845), (**b**) microstructured by MDC (q = 6.2 W/cm^2, Re = 1260), (**c**) covered by mesh (q = 5 W/cm^2, Re = 845) and (**d**) combined (q = 5 W/cm^2, Re = 1260) surfaces.

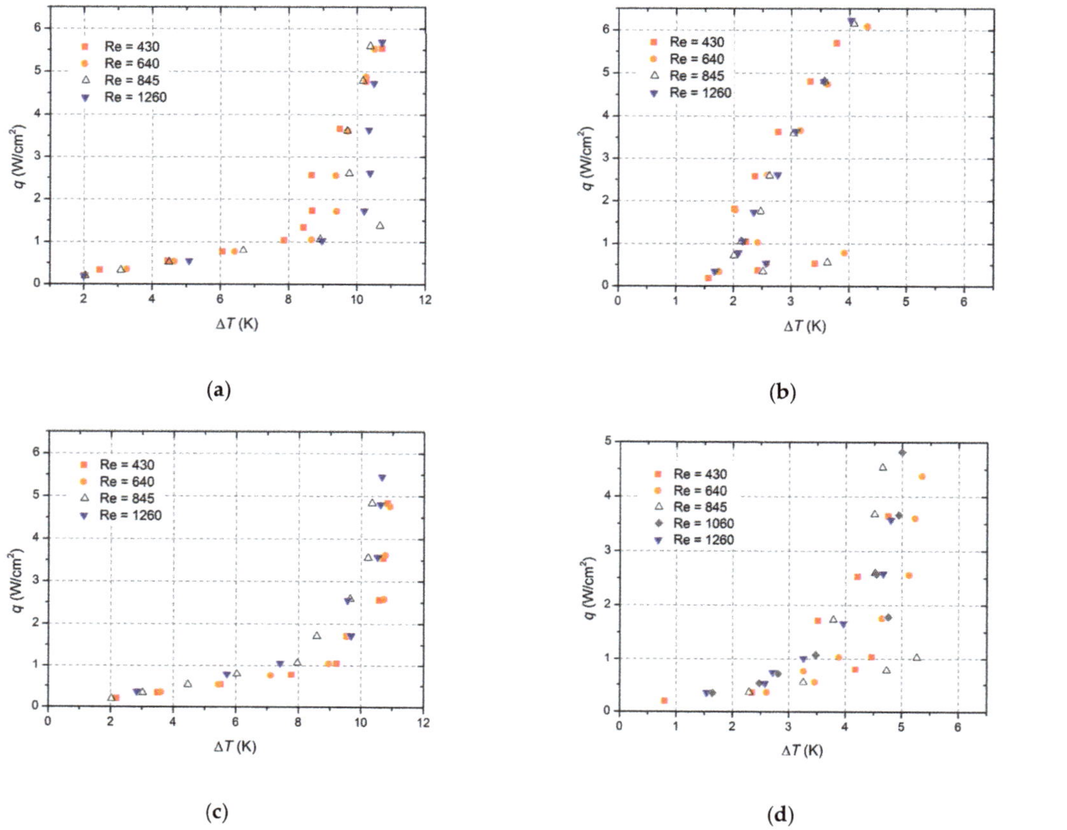

Figure 6. Boiling curves for: smooth (**a**), microstructured by MDC (**b**), smooth covered by mesh (**c**), and combined (**d**) surfaces.

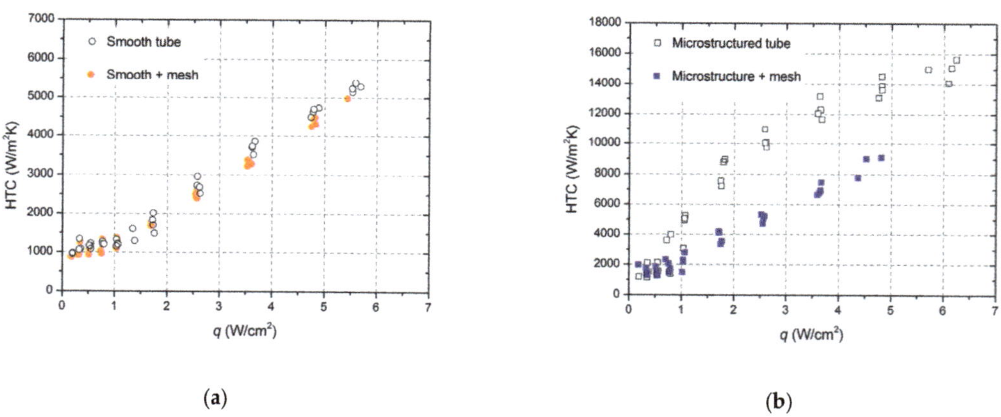

Figure 7. HTC vs. q dependencies for: smooth tube and smooth tube, covered by mesh (**a**); microstructured tube and microstructured tube, covered by mesh (**b**).

A brief discussion on the problem of the correct selection of the upper mesh covering parameters is given at the end of the section.

There are not many correlations for predicting pool boiling heat transfer coefficients for microstructured surfaces. We were able to find the calculation dependencies for predicting the pool boiling HTC for microfinned surfaces in [29,30]. Next, we tried to compare the obtained data for the MDC surface with these correlations, assuming that the process of developed nucleate boiling should be very similar for falling films and pool boiling. The main difference is brought by the evaporation of the film, however, in a developed boiling regime, its role decreases to a negligible value. We also neglected the presence of additional longitudinal knurling, present on the MDC-surface over the microfinning, since there are no calculation dependencies for such geometry in the literature (the correlation proposed in [29] for the prediction of boiling on micropin surfaces also gives greatly overestimated values in case of our MDC-surface).

The calculation dependence of Aksyanov et al. [29] has the form:

$$\alpha/\alpha_0 = 6K_q^{-0.2}\left(\frac{\theta}{90}\right)^{0.554}\left(\frac{h}{l_0}\right)^{0.19}\left(\frac{\Delta}{l_0}\right)^{0.201}(\delta/l_0)^{-0.394}, \quad (6)$$

where α, α_0 are the heat transfer coefficients of microstructured and smooth surfaces; K_q is the dimensionless criterion, namely, the scale of the averaged velocity of liquid resulting from a vapor generation process: $K_q = ql_0/(r\rho''\nu')$; l_0 is the Laplace constant $l_0 = \sqrt{\sigma/(g(\rho' - \rho''))}$; ρ' and ρ'' are the liquid and vapor density; ν' is the fluid kinematic viscosity; r is the latent heat of vaporization; σ is the surface tension coefficient. Geometrical parameters are the following: the angle of fin inclination—θ; fin height—h; gap between the fins—Δ; average fin thickness—δ.

The formula for α_0 expresses a standardized heat transfer coefficient:

$$\alpha_0 = 872P_{cr}^{\frac{1}{3}}/\left(T_{cr}^{\frac{5}{6}}M^{\frac{1}{6}}\right)\left(\frac{P}{P_{cr}}\right)^{0.1}\left(1 + 4.64\left(\frac{P}{P_{cr}}\right)^{1.16}\right)q^{2/3}, \quad (7)$$

where P_{cr} and T_{cr} are the critical pressure and critical temperature of the coolant, and M is the molecular weight of the coolant. The calculation dependence proposed in [29] by Formulas (6) and (7) makes it possible to predict heat transfer without requiring empirical parameters. However, it is worth noting that Formula (7) in our case gave overestimated values of α_0, so we used our own experimental data for smooth surface (Figure 6a) as reference values.

The dependence proposed by Huang [30] has the form of:

$$\alpha = 180\left(\frac{k_l}{l_0\sqrt{2}}\right)\left(\frac{ql_0\sqrt{2}}{k_l T_S}\right)^{0.36}\left(\frac{\rho'}{\rho''}\right)^{0.3}\left(\frac{P_{fin}}{l_0\sqrt{2}}\right)^{-0.2}\left(\frac{\Delta}{l_0\sqrt{2}}\right)^{-0.5}, \quad (8)$$

where k_l is the liquid thermal conductivity; T_S—saturation temperature and P_{fin} is fin pitch.

Comparison of received experimental data on HTC vs. heat flux for microstructured by MDC surface with the correlations for microfinned surfaces (6,8) shown in Figure 8. For reference, Figure 8 also shows the experimental data obtained for a smooth surface. As noted, the heat transfer enhancement on the microstructured tube is about 3–4 times more as compared to the smooth surface even at small heat fluxes, and ONB begins at heat flux values about 2 times less than those for the smooth surface. In fact, the maximal heat transfer enhancement up to 4 times takes place, namely, in the region of small heat fluxes (q: 1 ÷ 2.6 W/cm^2) because in this case, we compare the quite developed boiling on the MDC-surface with a highly underdeveloped boiling regime on the smooth surface.

The dependence (6) for a microfinned structure quite well describes the obtained experimental data for the microstructured MDC-surface, even better than within a deviation of 30%, as stated by the authors of [29]. Some excess of the experimental data over the predicted data (here the logarithmic fit of calculated points is implied) in the region of heat fluxes $q \leq 4$ W/cm^2 can be associated with the presence of additional knurling, which creates partly closed pores and intensifies heat transfer as compared to microfinned surfaces (according to our assumptions), so with the fact that we used the dependence (6) developed for a pool boiling to describe the falling films data: it is known that the contribution of falling film evaporation can give an additional increase in heat transfer as compared to pool boiling, especially at low and moderate heat fluxes, reaching 20–30%. The appearance of a step in the calculated HTC values by (6) in Figure 8 in the region of low heat fluxes ($1 \div 1.8\ W/cm^2$) is due to the presence of an underdeveloped boiling regime on the smooth surface with low HTCs, which are substituted into (6) giving low values of α.

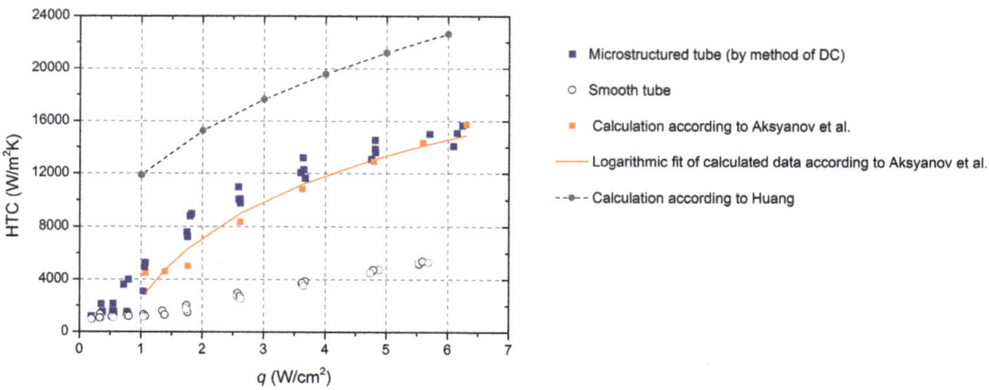

Figure 8. Comparison of HTC vs. q dependence for microstructured by deformational cutting tube with correlations for predicting pool boiling HTC for microfinned surfaces [29,30].

The dependence (8), on the contrary, does not work in our case, about two times exceeding the HTC data obtained, although it can be noted that both correlations have very similar character of the dependence of HTC on the heat flux.

In conclusion, it can be added that next exceeding the heat flux values provided in Figure 8 leads to the flattening of HTC vs. q dependence, associated with the appearance of washable dry spots, beginning from the lower part of the heated tube. Dry spots appear after reaching of CHF in places where the film is thinned, for example, such regions are visible in Figure 5a like bare zones (but in fact, these regions are still covered with a microlayer of liquid film). The microstructured surfaces, similar to those used in this work, in addition to increasing the heat transfer coefficient and reducing the ONB, provide the capillary replenishment of nucleation centers with liquid, making it possible to delay the development of the crisis phenomena. According to our estimates, based on a study of heat transfer on similar MDC-surfaces earlier [22], the minimal increase in CHF on the MDC-surface used in this work may be about 2 times as compared to the smooth surface, namely, to reach about $12\ W/cm^2$ for the upper values of the passed range of liquid flow rates.

Below, we briefly discuss the possible causes of reducing heat transfer when using the combined coverage.

As already mentioned we attribute the absence of heat transfer enhancement for the stainless steel (AISI 304) mesh-covered smooth copper surface primarily to the low thermal conductivity of the used mesh material ($\lambda \approx 15\ W/mK$) and the imperfect contact of the mesh with the heated surface (allowing the existence of gaps ~10 microns at the points of contact (despite our attempts to tighten the mesh covering on the tube as closely as possible). Maybe the second reason is even more important because there are works in which it is shown that steel coverings can be effective means of heat transfer. For example, in [31] it is shown that HTC provided by 3D-printed capillary-porous stainless steel coating (LPW 155, $\lambda \approx 20\ W/mK$) can be higher than HTC of the brass one (AISI C836000, $\lambda \approx 89\ W/mK$) with the same parameters of porous structure, at boiling of n-dodecane in horizontal liquid layers at reduced pressures. Thus, these issues require additional research.

Also, especially in the case of combined surfaces, we should not discount the possibility that guided by the choice of mesh parameters based on the empirical data of other authors [5,6,8] as well as our own previous data for microstructured surfaces [22], we may not choose the mesh coverage parameters well enough for our experimental conditions, which may cause the deterioration of vapor removal.

Studying the one-layer mesh coverings with an aperture size larger than the bubble departure diameter as well as the order of the departure diameter and smaller than the departure diameter, Tolubinskiy et al. [32] showed that the maximum HTC is realized at an aperture of the order of the bubble departure diameter. Based on this, we can assess that it is also true for the upper layer of two-layer coverings (or combined coatings with mesh covering the basic microstructure, like in our case). Then, the basic microstructure—or fine mesh in the case of two-layer mesh covering—is responsible for the intensification of the nucleation process and the upper coverage at least should not prevent effective vapor removal.

To estimate the value of departure diameter for the R114/R21 refrigerant mixture use the Labuntsov formula [33]:

$$D = 1.8\left(\frac{\sigma d_0}{g(\rho' - \rho'')}\right)^{1/3} \tag{9}$$

where d_0 is the characteristic size of the microroughness of the heating surface.

Based on Mahmoud and Karayiannis [34] calculation, who used the Hsu model [35] to predict the range of active cavity size for different liquids at atmospheric pressure and 5 K subcooling (5.5–126 μm, 0.7–108 μm, 0.4–79 μm—for water, HFE-7100 and FC-72, respectively), we can assume, without greatly deviating from the truth, that the range of active nucleation sites will be 1–100 μm for the R114/R21 refrigerant mixture (note, that this range includes the transverse dimensions of the gap between microfins, Figure 1b). Substituting the minimum and maximum values of the range into Formula (9) as d_0, we have: $D \approx 190$ μm for the smallest cavities and $D \approx 890$ μm for the biggest ones.

Thus, according to the calculation, by using the mesh covering with an aperture of 220 μm perhaps we are preventing vapor removal. When choosing the mesh parameters, besides basing the choice on our own results on enhancing surfaces microstructure parameters [22], we were guided by the parameters of enhancing covering from [6] (consisting of three upper layers of mesh with aperture 254 μm and diameter 50 μm) as well as by the uniform covering parameters from [8] (with aperture 100 μm and diameter 30 μm)—in both cases, the aperture of upper coverings was significantly less than the bubble departure diameter for water ($D \approx 2$ mm under standard conditions). As a result, our preference was given to mesh with an aperture of 220 μm (available with a wire diameter of 100 μm), slightly exceeding the pitch of the lower microfinning by MDC (200 μm). However, we did not take into account the fact that in previous work [6,8] boiling was carried out at high heat fluxes ~100 W/cm^2, an order of magnitude higher than maximum operating heat fluxes ~10 W/cm^2 in case of boiling refrigerant mixture R114/R21 films. At such high heat fluxes, cavities with the smallest dimensions can be activated. So, the different operating ranges of heat fluxes used in different works may be a possible cause of the discrepancy with respect to the choice of optimal aperture values of upper mesh coverings.

However, these are just some preliminary considerations, the issue of choosing covering geometrical parameters requires more precise analysis and calculations.

4. Conclusions

The experimental data on heat transfer coefficients at evaporation and boiling of a refrigerant mixture R114/R21 film falling over the smooth or microstructured surfaces and the same surfaces, covered by micromesh, are obtained. It is shown that:

- A microstructured surface created by deformational cutting demonstrates significant heat transfer enhancement—up to four times as compared to the smooth surface.

- Adding mesh covering with an aperture of 220 μm reduces the enhancement reached on the microstructured surface alone by up to two times. Thus, the heat transfer enhancement, provided by a combined surface, is about two times as compared to the smooth surface.
- The chosen mesh coverage overlaying the smooth surface does not give heat transfer enhancement compared to the smooth surface.
- The absence or deterioration of heat transfer enhancement on studied mesh-covered surfaces can be associated with the low thermal conductivity of used mesh material and especially the imperfectness of the mesh mounting method. The applied mesh covering possibly prevents the effective vapor removal during the boiling of the R114/R21 mixture falling film.
- The correlation, proposed by Aksyanov et al. for predicting the pool boiling heat transfer on microfinned surfaces, describes the obtained film flow data on heat transfer on the MDC surface quite well, while the correlation of Huang gives twice the values of the heat transfer coefficient.

Author Contributions: Conceptualization, O.V. and N.P.; methodology, O.V. and N.P.; validation, O.V. and N.P.; investigation, O.V. and N.P.; resources, A.P.; writing—original draft preparation, O.V.; writing—review and editing, O.V. and N.P.; visualization, O.V.; project administration, A.P.; funding acquisition, A.P. All authors have read and agreed to the published version of the manuscript.

Funding: This research was funded by a mega-grant from the RF Ministry of Science and Higher Education (No. 075-15-2021-575); studies of heat transfer on the MDC-surface were carried out within the framework of the state assignment of the IT SB RAS.

Data Availability Statement: Not applicable.

Conflicts of Interest: The authors declare no conflict of interest. The funders had no role in the design of the study; in the collection, analyses, or interpretation of data; in the writing of the manuscript; or in the decision to publish the results.

Nomenclature

A	heat releasing surface area
C	correction factor
D	bubble departure diameter
d	tube diameter; characteristic size of microroughness
G	mass flow rate
g	gravity acceleration
h	heat transfer coefficient; fin height
I	current through the heating element
k_l	liquid thermal conductivity
l_0	Laplace constant
P	Pressure
P_{fin}	fin pitch
q	heat flux density
R	heating element resistance
R_a	mean roughness
Re	film Reynolds number, $Re = \frac{4G}{\pi d \mu}$
T	temperature
U	uncertainty
Greek symbols	
α	heat transfer coefficient
Δ	gap between the fins
δ	liquid film thickness, average fin thickness
δ_{gap}	size of gap between fin tips

θ	the angle of fin inclination
λ	thermal conductivity
μ	dynamic viscosity
ρ	Density
σ	surface tension
Subscripts	
cr	Critical
s	Saturation
w	Wall
Superscripts	
′	liquid phase
″	vapor phase

References

1. Huang, Y.; Wang, M.; Xu, L.; Deng, J. Experimental study on a spray and falling-film cooling system. *Case Stud. Therm. Eng.* **2021**, *26*, 101057. [CrossRef]
2. Liang, G.; Mudawar, I. Review of pool boiling enhancement by surface modification. *Int. J. Heat Mass Transf.* **2019**, *128*, 892–933.
3. Dedov, A.V. A Review of Modern Methods for Enhancing Nucleate Boiling Heat Transfer. *Therm. Eng.* **2019**, *66*, 881–915. [CrossRef]
4. Volodin, O.A.; Pecherkin, N.I.; Pavlenko, A.N. Heat Transfer Enhancement at Boiling and Evaporation of Liquids on Modified Surfaces—A Review. *High Temp.* **2021**, *59*, 405–432. [CrossRef]
5. Volodin, O.A.; Pavlenko, A.N.; Pecherkin, N.I. Heat Transfer Enhancement on Multilayer Wire Mesh Coatings and Wire Mesh Coatings Combined with Other Surface Modifications—A Review. *J. Eng. Thermophys.* **2021**, *30*, 563–596. [CrossRef]
6. Zhang, S.; Jiang, X.; Li, Y.; Chen, G.; Sun, Y.; Tang, Y.; Pan, C. Extraordinary Boiling Enhancement Through Micro-Chimney Effects in Gradient Porous Micromeshes for High-Power Applications. *Energy Convers. Manag.* **2020**, *209*, 112665. [CrossRef]
7. Asakavičius, J.P.; Zukauskas, A.A.; Gaigalis, V.A.; Eva, V.K. Heat Transfer from Freon-113, Ethyl Alcohol and Water with Screen Wicks. *Heat Transf. Sov. Res.* **1979**, *11*, 92–100.
8. Tang, K.; Bai, J.; Chen, S.; Zhang, S.; Li, J.; Sun, Y.; Chen, G. Pool Boiling Performance of Multilayer Micromeshes for Commercial High-Power Cooling. *Micromachines* **2021**, *12*, 980. [CrossRef]
9. Sloan, A.; Penley, S.; Wirtz, R.A. Sub-Atmospheric Pressure Pool Boiling of Water on a Screen-Laminate Enhanced Surface. In Proceedings of the 25th Annual IEEE Semiconductor Thermal Measurement and Management Symposium, San Jose, CA, USA, 15–19 March 2009.
10. Tsay, J.Y.; Yan, Y.Y.; Lin, T.F. Enhancement of Pool Boiling Heat Transfer in a Horizontal Water Layer Through Surface Roughness and Screen Coverage. *Heat Mass Transf.* **1996**, *32*, 17–26. [CrossRef]
11. Brautsch, A.; Kew, P.A. The Effect of Surface Conditions on Boiling Heat Transfer from Mesh Wicks. In *International Heat Transfer Conference Digital Library*; Begel House Inc.: Danbury, CT, USA, 2002.
12. Chien, L.H.; Tsai, Y.L. An Experimental Study of Pool Boiling and Falling Film Vaporization on Horizontal Tubes in R-245fa. *Appl. Therm. Eng.* **2011**, *31*, 4044–4054. [CrossRef]
13. Chien, L.H.; Hwang, H.L. An Experimental Study of Boiling Heat Transfer Enhancement of Mesh-on-Fin Tubes. *J. Enhanc. Heat Transf.* **2012**, *19*, 75–86. [CrossRef]
14. Kim, H.; Park, Y.; Kim, H.; Lee, C.; Jerng, D.W.; Kim, D.E. Critical Heat Flux Enhancement by Single-Layered Metal Wire Mesh with Micro and Nano-Sized Pore Structures. *Int. J. Heat Mass Transf.* **2017**, *115*, 439–449. [CrossRef]
15. Dąbek, L.; Kapjor, A.; Orman, Ł.J. Distilled Water and Ethyl Alcohol Boiling Heat Transfer on Selected Meshed Surfaces. *Mech. Ind.* **2019**, *20*, 701. [CrossRef]
16. Zhang, C.; Zhang, L.; Xu, H.; Li, P.; Qian, B. Performance of Pool Boiling with 3D Grid Structure Manufactured by Selective Laser Melting Technique. *Int. J. Heat Mass Transf.* **2019**, *128*, 570–580. [CrossRef]
17. Pastuszko, R.; Kaniowski, R.; Wójcik, T.M. Comparison of Pool Boiling Performance for Plain Micro-Fins and Micro-Fins With a Porous Layer. *Appl. Therm. Eng.* **2020**, *166*, 114658. [CrossRef]
18. Huang, S.; Wang, L.; Pan, Z.; Zhou, Z. Experimental Investigation of a New Hybrid Structured Surface for Subcooled Flow Boiling Heat Transfer Enhancement. *Appl. Therm. Eng.* **2021**, *192*, 116929. [CrossRef]
19. Chen, G.; Fan, D.; Zhang, S.; Sun, Y.; Zhong, G.; Wang, Z.; Wan, Z.; Tang, Y. Wicking capability evaluation of multilayer composite micromesh wicks for ultrathin two-phase heat transfer devices. *Renew. Energy* **2021**, *163*, 921–929. [CrossRef]
20. Hu, Y.; Lei, Y.; Liu, X.; Yang, R. Heat transfer enhancement of spray cooling by copper micromesh surface. *Mater. Today Phys.* **2022**, *28*, 100857. [CrossRef]
21. Smakulski, P.; Sławomir, P. A review of the capabilities of high heat flux removal by porous materials, microchannels and spray cooling techniques. *Appl. Therm. Eng.* **2016**, *104*, 636–646. [CrossRef]
22. Volodin, O.; Pecherkin, N.; Pavlenko, A.; Zubkov, N. Surface Microstructures for Boiling and Evaporation Enhancement in Falling Films of Low-Viscosity Fluids. *Int. J. Heat Mass Transf.* **2020**, *55*, 119722. [CrossRef]

23. Zubkov, N.N.; Ovchinnikov, A.I.; Vasil'ev, S.G. Tool–workpiece interaction in deformational cutting. *Russ. Eng. Res.* **2016**, *36*, 209–212. [CrossRef]
24. Zubkov, N.; Poptsov, V.; Vasiliev, S. Surface Hardening by Turning without Chip Formation. *Jordan J. Mech. Ind. Eng.* **2017**, *11*, 13–19.
25. Pecherkin, N.I.; Pavlenko, A.N.; Volodin, O.A. Heat transfer and critical heat flux at evaporation and boiling in refrigerant mixture films falling down the tube with structured surfaces. *Int. J. Heat Mass Transf.* **2015**, *90*, 149–158. [CrossRef]
26. Zhukov, V.E.; Mezentseva, N.N.; Pavlenko, A.N. Teplootdacha Pri Kipenii Na Modifitsirovannoy Poverhnosti Vo Freone R21 I Smesi Freonov R114/R21 (Boiling Heat Transfer on Modified Surface in R21 Freon and in R114/R21 Freon Mixture). In Proceedings of the XXXVIII Siberian Thermophysics Seminar dedicated to the 65th Anniversary of the Kutateladze Institute of Thermophysics SB RAS, Novosibirsk, Russia, 29–31 August 2022; pp. 106–111. (In Russian).
27. Dang, C.; Jia, L.; Peng, Q.; Huang, Q.; Zhang, X. Experimental and analytical study on nucleate pool boiling heat transfer of R134a/R245fa zeotropic mixtures. *Int. J. Heat Mass Transf.* **2018**, *119*, 508–522. [CrossRef]
28. Shamirzaev, A. On the Pressure Drop Calculation During the Flow of Two-Phase Non-Azeotropic Mixtures. *Int. J. Multiph. Flow* **2022**, 104314. [CrossRef]
29. Aksyanov, R.A.; Kokhanova, Y.S.; Kuimov, E.S.; Gortyshov, Y.F.; Popov, I.A. Recommendations for Improving the Efficiency of Radio-Electronic Equipment Cooling Systems. *Russ. Aeronaut.* **2021**, *64*, 291–296. [CrossRef]
30. Huang, L.D. Pool Boiling Correlations for Structured Fin Tubes. In Proceedings of the 10th International Conference on Boiling and Condensation Heat Transfer, Nagasaki, Japan, 12–15 March 2018.
31. Shvetsov, D.A.; Pavlenko, A.N.; Brester, A.E.; Zhukov, V.I. Inversiya Crivoy Kipeniya Na Mikrostrukturirovannyh Poristyh Pokrytiyah (Boiling Curve Inversion on Microstructured Porous Coatings). In Proceedings of the 8th Russian national conference on heat transfer "PHKT-8", Moscow, Russia, 17–22 October 2022; pp. 83–84. (In Russian).
32. Tolubinskiy, V.I.; Antonenko, V.A.; Ivanenko, G.V. Crisis Phenomena in Boiling on Submerged Wire Mesh-Wrapped Wall. *Heat Transf. Sov. Res.* **1989**, *21*, 531–535.
33. Labuntsov, D.A. Heat Transfer and Vapor Dynamics. In *Current Concepts of the Mechanism of Nucleate Boiling of Liquids*; Institute of High Temperatures: Moscow, Russia, 1974; Available online: https://inis.iaea.org/collection/NCLCollectionStore/_Public/06/171/6171615.pdf (accessed on 28 November 2022).
34. Mahmoud, M.M.; Karayiannis, T.G. Pool boiling review: Part I–Fundamentals of boiling and relation to surface design. *Therm. Sci. Eng. Prog.* **2021**, *25*, 101024. [CrossRef]
35. Hsu, Y.Y. On the size range of active nucleation cavities on a heating surface. *J. Heat Transf.* **1962**, *84*, 207–213. [CrossRef]

energies

MDPI

Article

Nonisothermal Evaporation of Sessile Drops of Aqueous Solutions with Surfactant

Sergey Misyura [1,*], Andrey Semenov [1,2], Yulia Peschenyuk [1,2], Ivan Vozhakov [1,2] and Vladimir Morozov [1]

1 Kutateladze Institute of Thermophysics SB RAS, 630090 Novosibirsk, Russia
2 Faculty of Physics, Novosibirsk State University, 630090 Novosibirsk, Russia
* Correspondence: misura@itp.nsc.ru

Abstract: In recent decades, electronic devices have tended towards miniaturization, which necessitates the development of new cooling systems. Droplet cooling on a heated wall is effectively used in power devices with high heat flux densities. The use of a surfactant leads to an increase in the diameter of the wetted spot and the rate of droplet evaporation. Despite the wide interest and numerous works in this area, there are still unexplored questions regarding the influence of surfactant and wall temperature on convection, of nonisothermality, and of the decrease in the partial pressure of vapor with increasing surfactant concentration. This work experimentally studies the effect on the rate of droplet evaporation of wall temperature in the range 20–90 °C and of the concentration of surfactant in an aqueous solution of sodium lauryl sulfate (SLS) from 0 to 10,000 ppm. It is shown for the first time that an inversion of the evaporation rate related to the droplet diameter occurs with increasing wall temperature. The influence of key factors on the evaporation of a water droplet with SLS changes with temperature. Thus, at a slightly heated wall, the growth of the droplet diameter becomes predominant. At high heat flux, the role of nonisothermality is predominant. To determine the individual influence of the surfactant on the partial pressure of water vapor, experiments on the evaporation of a liquid layer were carried out. The obtained results and simplified estimates may be used to develop existing calculation models, as well as to optimize technologies for cooling highly heated surfaces.

Keywords: droplet evaporation; heated wall; surfactant; heat transfer; free convection

Citation: Misyura, S.; Semenov, A.; Peschenyuk, Y.; Vozhakov, I.; Morozov, V. Nonisothermal Evaporation of Sessile Drops of Aqueous Solutions with Surfactant. *Energies* **2023**, *16*, 843. https://doi.org/10.3390/en16020843

Academic Editor: Chi-Ming Lai

Received: 7 December 2022
Revised: 26 December 2022
Accepted: 29 December 2022
Published: 11 January 2023

1. Introduction

In the modern world, the main tasks of technical design, such as weight reduction and object miniaturization, face implementation problems due to the disadvantages of conventional heat removal systems. Efficient dissipation of high heat fluxes requires new approaches to cooling [1]. At present, various two-phase heat exchangers such as spray systems [2], film cooling [3], and microchannel heat exchangers [4] are being widely investigated. In addition, some researchers are considering the possibility of changing the properties of the working fluid with the help of various additives [5,6].

One of the new directions in the study of the effectiveness of cooling systems is the use of aqueous solutions with surfactants as a working fluid [7]. However, the addition of a surfactant can reduce the liquid evaporation rate, leading to a decrease in the equipment efficiency. The easiest way to find out the influence of this factor is to study the evaporation of an immobile sessile drop. The problem of an evaporating drop has attracted the attention of researchers since the end of the last century. A well-known problem is that of the "coffee drop" [8]: the drop is deposited on a rough substrate and its contact line remains stationary during evaporation, when the capillary flow carries the material from the center to the edge and, on drying, results in a circular trail. Another case is when the drops are applied to a smooth, well-wettable surface. In this case, the drop spreads, i.e., the contact line is movable [9,10].

The theoretical description of evaporating droplets is complicated by a large number of physical effects that must be taken into account. The problem of droplet spreading and evaporation is the subject of many works, which are summarized in a sound review by Bonn et al. [11]. In the classical mathematical formulation of the mass transfer model in single-component spherical droplets evaporating in still air, Maxwell's approach was used [12]. It was assumed that the ratio of diffusion time to evaporation time $R_d^2/D_v\tau_e \approx \rho_{vs}(1-RH)/\rho_a$ is small, and then evaporation may be considered as a quasi-stationary process. In this case, the mass flow is limited by diffusion and for a spherical drop, it is equal to:

$$J_M = 4\pi R_d D_v(\rho_{vs} - \rho_{va})$$

Assuming that the vapor obeys the ideal gas law, the vapor density can be expressed in terms of the partial pressure:

$$\rho_v = \frac{p_v M_v}{RT} = \frac{RHP_{sat}(T)M_v}{RT}$$

In the case of evaporation of a liquid drop in air, the resulting vapor on the droplet surface displaces the surrounding vapor-air mixture and forces it to move away from the drop at a certain velocity. This phenomenon is called the Stefan flow. The evaporation rate of a spherical droplet is limited by the Stefan flow [13]:

$$J_S = 4\pi R_d D_v \rho_a ln(1+B_M)$$
$$B_M = (\rho_{vs} - \rho_{va})/(\rho_a - \rho_{vs})$$

Tonini and Cossali [14] extended this model to the case of nonisothermal evaporation of a spherical drop in the surrounding atmosphere.

When considering small sessile drops with a contact radius less than the capillary length $\lambda_c = \sqrt{\sigma/\rho_l g}$, the gravitational forces are insignificant, so the liquid takes the form of a spherical cap. The height h_d, surface area S_d and volume V_d of a spherical cap can be calculated using the wet spot radius R_d and the contact angle θ:

$$h_d = R_d tan(\theta/2)$$
$$S_d = \pi(h_d^2 + R_d^2)$$
$$V_d = \tfrac{\pi h_d}{6}(3R_d^2 + h_d^2)$$

Hu and Larsen [15] developed an isothermal evaporation model for a sessile drop. An analytical dependence of the mass flow on the contact angle was obtained. The developed theory is in good agreement with the experimental data and the results of numerical calculations for contact angles from 0 to $\pi/2$. According to this model, the mass flow is:

$$J_H = \pi R_d D_v(\rho_{vs} - \rho_{va})(0.27\theta^2 + 1.3)$$

Experimental studies of the evaporation of a sessile drop have been carried out for a long time and have resulted in extensive and versatile data. The papers [16–18] deal with water drop evaporation in the open air, showing the influence of such parameters as substrate material, angle of inclination, and environmental conditions on the evaporation process. By now, a relatively large amount of information on the evaporation of various liquids has been collected.

Tarasevich [19] performed an analytical analysis of hydrodynamics inside a sessile drop at the initial stage of evaporation. The qualitative picture of the velocity field was shown to be independent of the ratio between the drop height and the radius of the contact line. Saada et al. [20] numerically simulated the evaporation of a drop lying on a heated substrate. The simulation took into account the convection of the surrounding air as a result of heating from the substrate, but heat was considered to be transferred inside the drop only due to thermal conduction. The results revealed an underestimation of the evaporation rate compared to the experimental data, and the error increased with increasing substrate temperature. Timm et al. [21] presented data on the evaporation rate in comparison with the diffusion dependence. This assumption was shown to be valid when the time scale of vapor concentration diffusion was much smaller than the total time of droplet evaporation. However, these conditions are valid for experiments in open air.

Misyura [22] demonstrated that at high corrosion rates and intense gas evolution, the evaporation rate decreases due to a decrease in the rate of heat supply to the droplet surface. Modeling the corrosion kinetics also requires taking into account the statistical nature of the corrosion process, which depends on the wettability and droplet diameter. It was shown for the first time that the corrosion kinetics is determined by the wetting regimes. Misyura et al. [23] also studied the effect of the thermocapillary convection and surfactant in a sessile droplet during nonisothermal evaporation.

Misyura [24] investigated the influence of the free convection. The Marangoni thermal convection was shown to play a predominant role in the heat transfer in the liquid evaporation.

Truskett and Stebe [25] investigated the effect of an insoluble surfactant monolayer on the evaporation rate. In this work, no difference from pure water was found. Kim et al. [26] studied the flash evaporation of water droplets with the addition of a surfactant. The latter was shown to strongly affect the shape of the evaporating droplet due to Marangoni forces. Semenov et al. [27] conducted an experimental and theoretical study of the contact angle during the evaporation of drops with surfactant solutions. Guti'errez et al. [28] proved that during vacuum evaporation, a surfactant reduces the rate of water evaporation.

For normal liquids, in particular pure water and most aqueous surfactant solutions [29], the surface tension decreases with increasing temperature. This means that the derivative of the surface tension coefficient with respect to temperature is negative. Hence, thermocapillary Marangoni forces arise with a nonuniform temperature distribution on the surface. The result is an outflow of fluid from warm area to cold area.

The term "self-wetting" was introduced by Abe et al. [30], who studied the thermophysical properties of dilute aqueous solutions of high-carbon alcohols. Due to thermocapillary stresses and the shape of the surface tension–temperature curve, the investigated liquids spread in a "self-wetting" manner, spontaneously propagating towards hot areas, and thereby preventing hot surfaces from drying out and increasing the rate of heat transfer. Because of these properties, "self-wetting" liquids were associated with significantly higher critical heat fluxes compared to water [31]. Savino et al. studied the behavior of vapor locks inside wickless heat pipes [32]. They found that the plugs were significantly smaller than those of liquids such as water. The study of self-wetting liquids was carried out in microgravity for space applications on the International Space Station [33]. Hu et al. [34] demonstrated that the use of these fluids in micro-oscillatory heat pipes resulted in an increase in heat pipe efficiency.

The use of alcoholic solutions with water may be undesirable, since they can significantly differ in thermophysical properties from pure water. At the same time, a very small amount of this substance is used for creating a surfactant solution, which does not affect its thermophysical properties, but noticeably changes its surface properties.

The performed analysis of existing works has shown that there are very few experimental and theoretical data on the influence of various factors on the evaporation of a drop of an aqueous solution with surfactant in a wide range of wall temperatures. The complexity of these studies is related to the fact that at high-temperature evaporation, the isothermal model of diffusion is unacceptable and evaporation is influenced simultaneously by many key factors. In addition, the properties of the surfactant change strongly with increasing liquid temperature.

Previous experimental studies have demonstrated that the addition of a surfactant leads to a decrease in the evaporation rate of a liquid droplet. However, the specific factors influenced by the surfactant have not been established. In addition, there has been no comprehensive study of the effect of the surfactant in a wide range of wall temperatures and droplet base diameters.

Research objectives of this article are as follows. (1) Experimental determination of geometrical parameters of a droplet and measurement of temperature fields of a liquid droplet. The obtained experimental data are used to predict the evaporative behavior of droplets in a wide range of wall temperatures. (2) On the basis of experimental data, approximate calculations are made to estimate the influence of separate key factors on the droplet evaporation rate. The findings may be useful for the development of the existing models for determining the evaporation rate of droplets of solutions with surfactants, which is important for the creation of droplet irrigation technologies, widely used for heat exchange intensification in devices with high energy efficiency.

2. Experimental Setup

Preparation for the experimental study of droplet evaporation involved obtaining aqueous solutions with surfactants. In the presented work, three different surfactants were selected: polyoxyethylene sorbitan monooleate (TWEEN 80), cetrimonium bromide (CTAB), and sodium lauryl sulfate (SLS). They differed in properties depending on the charge component of their hydrophilic components: nonionic (TWIN80), cationic (CTAB), or anionic (SLS).

Using a high-precision micro analytical balance AND BM-252 with a readability of 0.01 mg, the weight of the surfactant was measured. It was then mixed with deionized nanofiltered Milli-Q pure water, and the solution was thoroughly stirred at room temperature in a closed cuvette, using a magnetic stirrer with medium speed.

Attempts to obtain an aqueous solution of TWIN80 have demonstrated that producing the solutions of nonionic surfactants requires a special procedure, described in chemical protocols, including special conditions and additional reagents. Concluding that TWIN80 is poorly soluble in water, it was decided not to use it for our experiments. The study of the remaining solutions is carried out using two experimental setups described below.

2.1. Tensiometer KRÜSS K100

The analysis of the properties of aqueous solutions with surfactants was carried out on a precision tensiometer KRÜSS K100 with KRÜSS laboratory software. Using a Huber thermostat, a cuvette with a working liquid was heated or cooled in the temperature range from 294 K to 343 K. Measurements were performed using the Wilhelmy plate method. A standard 10 mm × 20 mm × 0.2 mm platinum plate was used to measure surface tension since the contact angle of such a plate with most liquids is approximately zero. The second plate with dimensions of 10 mm × 20 mm × 2.8 mm was made of copper of industrial grade A1 and was used to measure contact angles. The copper surface was technically treated. In most cases, copper with high thermal conductivity and low cost is used as a main material for heat pipes. In the case of using a nonstandard plate, a special holder was used to ensure a strictly vertical position for the copper plate.

The tensiometer placed the platinum strip in the solution, then slowly removed it at a constant rate until it was completely out of the solution. During this procedure, the tensiometer measured the force f with which the plate was removed. Furthermore, the value of the surface tension coefficient σ was calculated by the formula: $\sigma = f/(l \cdot cos\theta)$.

In this case, the contact angle θ between the platinum plate and the investigated fluid was taken to be zero. To measure the value of the contact angle with the copper plate, a similar procedure was followed. To verify the obtained data, measurements were carried out in series of 5 runs. As a result, statistically averaged data for the surface tension coefficient $\sigma(T)$ and the contact angle θ were obtained.

2.2. Drop Shape Analyzer KRÜSS DSA100

Experiments for the sessile liquid drop evaporation were performed with the use of the DSA100 drop shape analysis system, produced by KRÜSS and presented in Figure 1.

This device consists of three main parts: a high-precision dosing system with the dosing step of 0.1 μL, a motorized object table with program-driven movement in 2 horizontal axes, and the optical shadow system which includes a 50 W light source and a CCD camera with a resolution of 780 × 580 pixels (field of view varied from 3.7 mm × 2.7 mm to 23.2 mm × 17.2 mm). A Peltier chamber open to the atmosphere was used for heating substrates. It allowed the specified temperature of the bottom wall to be maintained with an accuracy of about 0.1 °C. The thermostat was used to cool the chamber and maintain the temperature of the substrate. The needle with the selected solution was moved into the chamber and a drop was deposited onto the substrate surface.

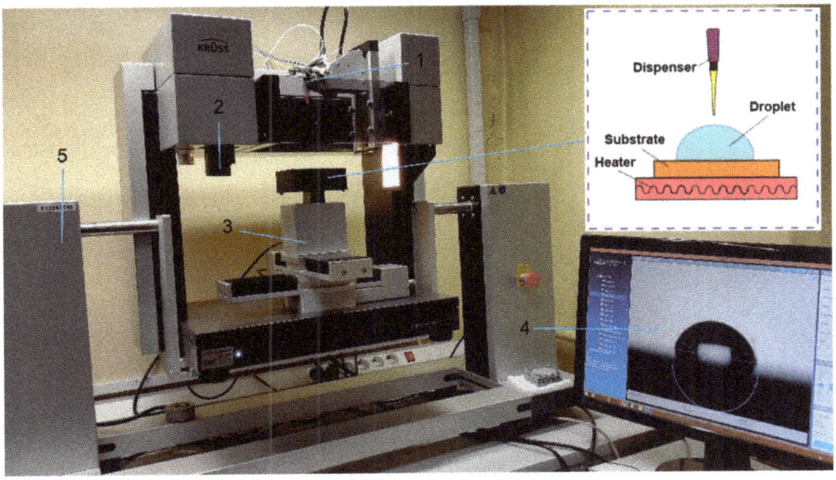

Figure 1. KRÜSS DSA 100 experimental setup: 1—dosing system; 2—camera; 3—three-axis positioning system; 4—processed image of a droplet; 5—automatic inclination system.

Measurement of the surface temperature of the copper substrate was carried out using a platinum resistance thermometer, fixed at the distance of 1–2 mm from the liquid drop. A special platinum resistance thermometer pt100 with a diameter of 2.5 mm, produced by KRÜSS, was applied. The accuracy of pt100 was 1/3 DIN class B (±0.1 °C at 0 °C to 0.8 °C at 400 °C). To increase the reliability of temperature measurements, the wall temperature was also measured by a thermocouple, located at a distance of 0.3–0.5 mm from the surface of the copper substrate on which a droplet was placed. The measurement error for the wall temperature (T_w) made by this thermocouple did not exceed 0.5–1 °C (taking into account the estimated distance of the thermocouple from the wall). The substrate surface temperature in each experiment was kept constant. The temperature and relative humidity of the ambient air during the experiments was 20–21 °C and 30–32%, respectively.

The measurement error of the droplet diameter did not exceed 5%. In the range of contact angle of 90–20°, the inaccuracy of the contact angle of the drop ranged within 3–5%. The maximum error of the evaporation rate did not exceed 5–7%.

The initial liquid drop volume was about 2–3 μL. The KRÜSSDSA Advance software allowed the processing of the shadow images of evaporating liquid droplets in fully automatic mode 2 (real-time determination of the droplet volume, base diameter and contact angles). Figure 2 shows a photograph of the liquid droplet shape analysis.

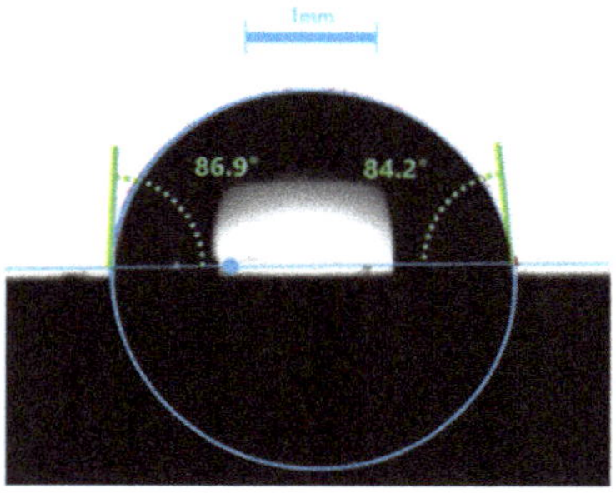

Figure 2. Photograph of liquid droplet shape analysis.

3. Experimental Results

The presented results of the experimental studies are divided into two main parts. To begin with, the analysis of aqueous solutions of SLS and CTAB on the KRÜSS tensiometer is considered. The second part is devoted to the results of the study of droplet evaporation.

3.1. *Surface Tension*

The wide temperature range specific for the operation of a pulsating heat pipe imposes restrictions on the applicability of working fluids. One of the determining roles in heat transfer and efficient operation of a pulsating heat pipe is played by the properties of the working fluid used.

At the outset, researchers obtained data on the dependence of surface tension on the concentration of surfactants in an aqueous solution at a room temperature of about 20 °C. For this purpose, a wide range of concentrations of surfactant solutions was used. The maximum concentration exceeded the CMC (critical micelle concentration), while the minimum concentration was similar in terms of surface tension to that of pure water. The components of the experimental setup were carefully prepared before each measurement. To reduce contamination caused by the surfactants used, the cuvette for the investigated working fluid was cleaned with a 70% isopropanol solution and washed several times with clean water.

Aqueous solutions with four different concentrations of SLS were investigated and pure water was examined for comparison. Figure 3 shows temperature dependence of surface tension of aqueous solutions with surfactants.

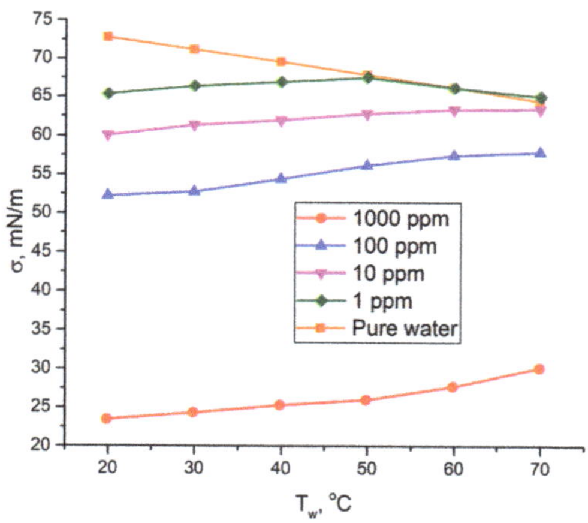

Figure 3. Surface tension vs. temperature for different concentrations of SLS in aqueous solutions.

The general dependence $\sigma(T)$ for CTAB is similar to that for water. That is, the first derivative of surface tension with respect to temperature is less than zero. In a heat exchange facility, for example, such as a pulsating heat pipe, liquid flows from the more heated area to the less heated one occurring on the surface due to Marangoni forces. This leads to the formation of dry spots and a decrease in the heat transfer coefficient, which may lead to overheating of the equipment. However, the SLS solution does not exhibit regular dependence, as CTAB, water and other normal liquids do [35]. Figure 3 shows that $d\sigma/dT$ is positive over a wide range of SLS concentrations. At the same time, it should be noted that at low SLS concentrations, a rapid change in the sign of the derivative is observed with increasing temperature. The temperature of the break point depends on the concentration. At higher temperature, the surface tension of the solution completely coincides with that of water.

Thus, aqueous solutions of SLS have the properties of a self-wetting liquid in a wide range of concentrations and temperatures. However, self-wetting liquids are quite often known to be unstable, which is also confirmed by the results of the current study.

From the obtained data, it may be inferred that SLS solutions are useful in solving the problem of dry spots and expanding the working range into the region of higher heat fluxes—with the proviso, of course, that for each task it is necessary to pay attention to the permissible range of applicability for each specific concentration of the solution.

3.2. Evaporation

In droplet evaporation, it is important to take into account the contact angle, since this determines the shape of the droplet surface. Therefore, the contact angles of evaporating droplets were measured for different concentrations of surfactant solutions and different substrate temperatures.

Figure 4 shows the evolution of a pure water drop during evaporation under isothermal conditions at a room temperature of about 21 °C. Since the drop has a small volume of the order of 1 µL, its shape is a spherical cap. In this case, evaporation apparently occurs at a fixed contact line, with the so-called "pinning" of the contact line. The evaporation is as a result of the diffusion of vapor from the drop surface, where it is saturated, into the atmosphere with a constant vapor concentration (in the experimental conditions, the humidity was about 30–40%). The evaporation is almost uniform over the entire surface area and is well described by Stefan's theoretical model [13]. Droplet evaporation on the heated surface qualitatively coincides with isothermal evaporation, but there are important differences. In this case, the mass flow from the droplet surface is substantially nonuniform. A significant mass flow is provided by intense evaporation in the area of the contact line, called the microregion [36].

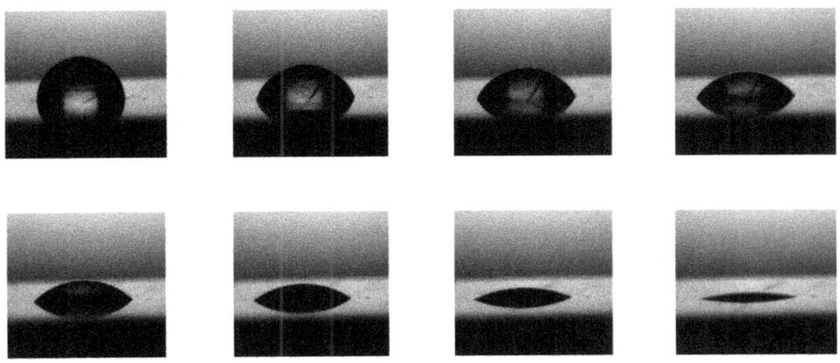

Figure 4. Evolution of an evaporating water droplet at a temperature of 21 °C (the initial droplet radius R_d = 0.7mm, the initial contact angle θ_0 = 97°). Time step equals 300 s.

In a further step, the evaporation of drops of surfactant solutions was considered in order to find out the effect of the surfactant concentration on the evaporation rate (heat exchange with the substrate) and the behavior of the contact line during evaporation.

The data for a small set of drops are presented below, whereby all the given graphs of diameter and contact angle correspond to the same drops.

Figures 5 and 6 show values of the droplet contact angle and the droplet height versus time during evaporation. At the initial moment of time, the contact angle of droplets of surfactant solutions is in the range from 35° to 100°. The initial droplet height is 0.4–1 mm. Upon evaporation, the contact angle decreases, which is associated with a change in the droplet volume, but the contact line remains almost immovable. It is clearly seen that pure water corresponds to the maximum contact angle for both low (30 °C) and higher (80 °C) temperatures.

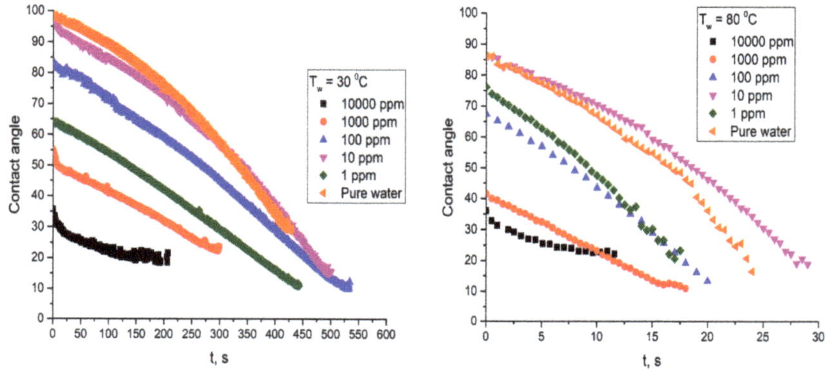

Figure 5. Evolution of contact angle over time of evaporation.

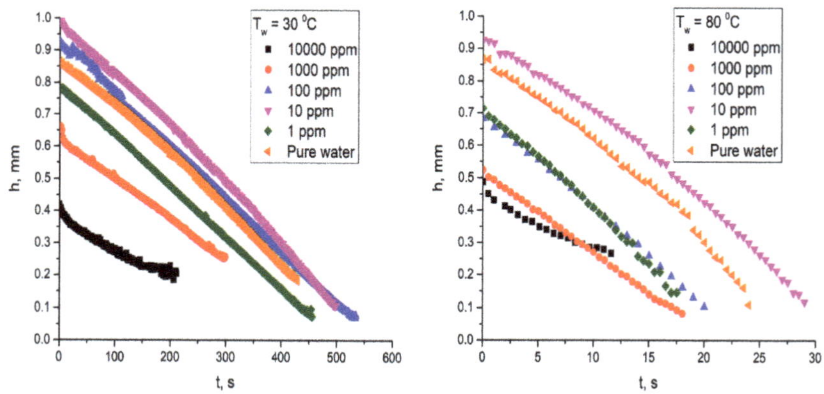

Figure 6. Evolution of drop height over time of evaporation.

Meanwhile, at a low surfactant concentration, the contact angle does not differ from that of pure water, and with an increase in the surfactant concentration, it monotonically decreases. There are two reasons for this. First, the surface tension reduction leads to a decrease in capillary forces and spreading of the droplet. Secondly, at the initial stage, the temperature of the liquid is close to room temperature and, upon contact with the heated substrate, it begins to heat up rapidly. It is supposed that the drop heating in the vicinity of the contact line occurs faster than in the main volume of the drop, which gives rise to the Marangoni force. The force vector is directed along the surface from lower to higher temperature. Estimating the time required to heat the drop to the substrate temperature, previous researchers obtained the formula $\tau = R^2\rho c_p/\lambda \approx 6$ s. Thus, the droplet diameter should grow noticeably in the first seconds of heating.

Figure 7 shows the droplet diameters as a function of time. At low surfactant concentrations in pure water, the diameter is seen to be constant during the entire evaporation time, except for the last seconds of the droplet life.

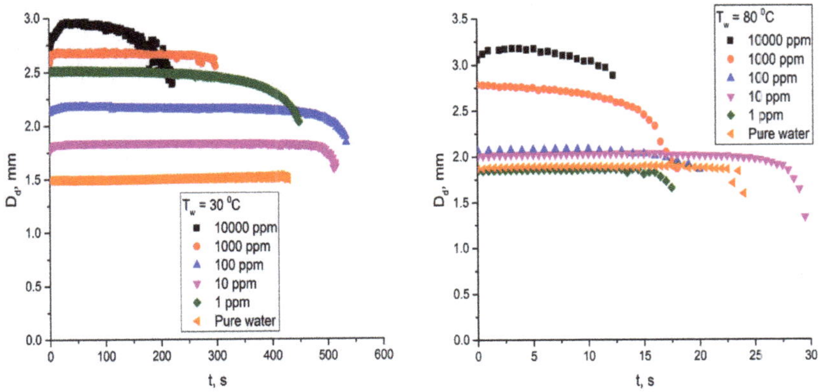

Figure 7. Evolution of dropletsdiameter over time of evaporation.

With increasing concentration, the area where the change in diameter occurs is clearly observed to stretch over time. Moreover, for a high concentration (above CMC), the droplet diameter changes throughout the entire evaporation process: in the first seconds, the drop diameter increases as the drop itself spreads, and then decreases as it dries.

4. Predicting the Evaporative Behavior of a Water Droplet with a Surfactant

Direct numerical simulation of the unsteady and nonisothermal evaporation of a water droplet with a surfactant is very complicated due to the large number of uncertain factors that affect each other. In the presence of a surfactant, it is hard to estimate the equilibrium diameter of the droplet, which can change markedly over time when the droplet evaporates. Any change in the diameter, contact angle or height of the droplet alters free convection in the gas and liquid phase. The surfactant affects the partial pressure of water vapor and, accordingly, the evaporation rate. To date, there are no reliable methods for computing all these factors. Calculating the temperature of the free surface of the droplet is extremely difficult, since the Marangoni flow and convection in a droplet depend on the surfactant concentration. Therefore, assessment of the influence of various key factors on the evaporative behavior of the droplet becomes crucial at the first stage. For this purpose, it is advisable to use experimental data on the geometry of the droplet, as well as on the thermal field measured by a thermal imager (T_s) and a thermocouple (T_w).

4.1. Key Factors Effecting the Droplet Evaporation

In general, the rate of evaporation depends on the geometry of the droplet, the conditions of heat transfer and on free convection. It is possible to distinguish the following key parameters that determine the intensity of evaporation of a water droplet with the addition of a surfactant:

(1) Drop diameter;
(2) Natural convection of gas;
(3) Natural convection of liquid;
(4) Nonisothermality;
(5) Buoyancy of water vapor in air and the Stefan flow;
(6) Evaporation rate of surfactant solution.

With a small droplet diameter (1–3 mm) and in the absence of wall or air heating, convection in the liquid and gaseous phases can be neglected. When a droplet evaporates on a hot wall, free convection has to be taken into account. The influence of these key factors on the evaporation of liquid are considered separately below.

(1) Let us take a detailed look at the influence of the wetted droplet diameter on the evaporation rate. In most experimental curves, there is an attached contact line for most of the evaporation time (evaporation mode of constant contact radius, or CCR). With the growth of the surfactant concentration, the droplet radius increases. Despite the decrease in the contact angle with time, for most of the curves, a quasi-linear decrease in the droplet volume is realized (i.e., at a fixed radius, the evaporation rate (dV/dt) is constant, despite a significant decrease in θ). Thus, the effect of the surfactant on evaporation due to the change in radius will be proportional to the ratio of the radii of the two drops ($J_1/J_2 \sim R_{d1}/R_{d2}$). To analyze the influence of other key factors, it is necessary to exclude the influence of the radius being a part of the ratio J/R_d.

(2) Let us estimate the effect of gas convection on the evaporation of a drop. When a liquid layer evaporates, the convective mass flow of water vapor through air can be estimated using Equation (1) [37],

$$Sh = 2 + 0.55(Re)^{0.5}(Sc)^{0.33} \tag{1}$$

where the Sherwood number $Sh = \beta R_d/D_v$, the Schmidt number $Sc = \nu_g/D_v$, the Reynolds number $Re = U_{ca}R_d/\nu_g$, in which β is the mass transfer coefficient, R_d is the radius of the droplet, D_v is the diffusion coefficient of water vapor in air, U_{ca} is the rate of thermogravitational convection of air at the temperature difference (ΔT_s) between the temperature of the free surface of the droplet T_s and the temperature of the ambient air T_a, and ν_g is the kinematic viscosity of the gas. The evaporation rate J is related to the mass transfer coefficient through the concentration difference $\rho_{vs} - \rho_{\text{va}}$ given in Equation (2),

$$J = \beta F(\rho_{vs} - \rho_{va}) \tag{2}$$

where F is the interface area of the layer.

For intense convection, when the Sherwood number is much greater than 1, $Sh = 0.55(Re)^{0.5}(Sc)^{0.33}$, then in accordance with Equations (1) and (2),

$$J \sim \beta \sim (U_{ca})^{0.5} \tag{3}$$

The air velocity due to thermogravitational convection can be approximated from Equation (4),

$$U_{ca} \sim (g\beta\Delta T_s R)^{0.5} \tag{4}$$

and the evaporation rate is related to the droplet radius according to Equation (5):

$$J \sim (R_d)^{0.5} \tag{5}$$

As indicated in Equation (5), the presence of the surfactant leads to an increase in the droplet radius; and an increase in the radius contributes to an increase in evaporation.

(3, 4) Let us evaluate the joint influence of factors 3 and 4, since they are interrelated. For convenience of analysis, the effect of these factors will be denoted as the effect of nonisothermality. The heat flux in the liquid q_l is supplied to the interface of the drop and spent on evaporation ($q_e = rJ$, $J = \Delta m/\Delta t$), radiation ($q_r = \varepsilon\sigma(T^4{}_s - T^4{}_a)$) and free air convection ($q_c = \alpha_a(T_s - T_a)$). The heat balance for the droplet surface corresponds to Equation (6),

$$q_l = q_e + q_r + q_c = rJ/F + \varepsilon\sigma\left(T^4{}_s - T^4{}_a\right) + \alpha_a(T_s - T_a) \tag{6}$$

where r is the latent heat of vaporization, and F is the area of the liquid interface. The air heat transfer coefficient α_a is determined by the Nusselt number $Nu = \alpha_a R_d/\lambda_a = 0.54\ Ra^{0.2}$, where Ra is the Rayleigh number [37]. The heat supplied to the liquid from the wall is determined by the conductive and convective heat transfer in the liquid according to Equation (7) [38–40],

$$\begin{gathered} q_l = \alpha_l(T_w - T_s) \\ \alpha_l = \frac{\lambda_l}{h}\left(\frac{U_{cl}h}{a_l}+1\right)^{0.3} \sim \frac{\lambda_l}{h}\left(\frac{kh^2}{a_l}+1\right)^{0.3} \sim \frac{1}{h^{0.4}},\ \text{if } \frac{U_{cl}h}{a_l} \gg 1 \\ \alpha_l = \frac{\lambda_l}{h}\left(\frac{U_{cl}h}{a_l}+1\right)^{0.3} \sim \frac{\lambda_l}{h}\left(\frac{kh^2}{a_l}+1\right)^{0.3} \sim \frac{1}{h},\ \text{if } \frac{U_{cl}h}{a_l} \ll 1 \end{gathered} \tag{7}$$

where α_l is the liquid heat transfer coefficient, $U_{cl} \sim c(Ma + Ra)$, U_l is the average convective velocity in the liquid droplet, the Marangoni number $Ma = (\Delta T_w h/\mu_l a_l)(d\sigma/dT_s)$, the Rayleigh number $Ra = g\beta_T \Delta T_w h^3/\nu_l a_l$, β_T is the thermal liquid expansion, g is the gravity acceleration, $\Delta T_w = T_w - T_s$, ν_l is the water kinematic viscosity, a_l is the liquid thermal diffusivity, and μL is the water dynamic viscosity. Empirical expressions for evaporation intensification with an increase in the Ra number are given in [41]. For very small droplets (droplet height below 1 mm), low values of Ra and Ma and low values of U_l are realized, and the condition $\alpha_l \sim 1/h$ (conductive heat transfer) is satisfied. Therefore, the ratio of heat fluxes for a pure water drop (q_{l1}) to a water drop with surfactant (q_{l2}) will be inversely proportional to the ratio of heights $q_{l1}/q_{l2} \sim h_2/h_1$. To determine the relationship between ΔT_w and the heat flux, it is necessary to solve a system of differential equations and it is difficult to make a simplified estimate for the change in $\Delta\rho_s$ due to nonisothermality. Therefore, the effect of nonisothermality can only be determined after the evaluation of other parameters (key factors 1, 2, 5, 6) and taking into account experimental data. At different wall temperatures, the droplet free surface temperature and ρ_s were determined using thermal imaging measurements and the equilibrium curve for the equilibrium partial pressure of water vapor.

(5) Let us consider the effect of water vapor buoyancy. Since water vapor is lighter than air, the effect of the additional flow must be considered. The vapor convection velocity U_ρ corresponds to Equations (8) and (9) [42],

$$U_\rho = ((\rho_a - \rho_m)gL/\rho_m)^{0.5} \tag{8}$$

$$\rho_m = \rho_a - (\rho_a - \rho_v)/\rho_v)C_{cat}(T_a) \tag{9}$$

where ρ_a is the density of the air, ρ_v is the density of the vapor, ρ_m is the density of the air + vapor mixture, and L is the characteristic length of the diffusion vapor layer. Experimental and calculated data in [42] show that the increase in the evaporation rate due to the buoyancy of water vapor (for a wall with high thermal conductivity) is about 40%. It does not depend on the droplet radius ($R_d > 0.8$ mm). Thus, despite the change in droplet diameter due to the surfactant, the effect of buoyancy on J will be the same for a water droplet with and without surfactant.

(6) Let us consider the influence of the surfactant on the droplet evaporation rate. Since the surfactant alters the radius of the drop, the simultaneous influence of several factors appears. To eliminate the effect of the droplet radius, one may consider the evaporation of a water layer (with and without surfactant). According to Equation (10),

$$J = \pi R_d D_v(\rho_{vs} - \rho_{va})(0.27\theta^2 + 1.3) \tag{10}$$

at zero contact angle, the evaporation rate does not depend on the angle: in this case, the expression for evaporation of the liquid layer thus coincides with the expression for a droplet with a very small height, Equation (11):

$$J = 4.1R_d D_v(\rho_{vs} - \rho_{va}) \tag{11}$$

4.2. Evaporation of a Layer of Water with Surfactant

Figure 8 presents experimental data on the evaporation of a layer of water with (10,000 ppm) and without surfactant. Figure 8a,b gives experimental curves for layer evaporation of T_w = 21 °C, and T_w = 90 °C, respectively. All curves demonstrate a linear dependence of the layer mass on time.

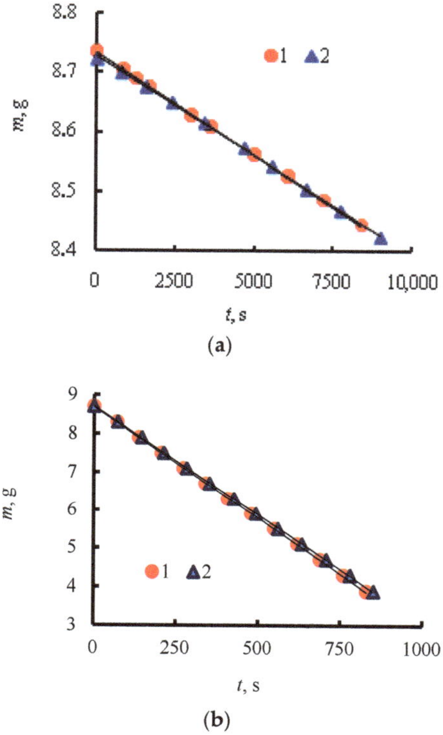

Figure 8. Evaporation of water layer (curve 1) and water layer with 10,000 ppm surfactant (curve 2) for layer diameter 40 mm and air humidity 30%: (**a**) $T_w = T_a$ = 21 °C, (**b**) T_w = 90 °C, T_a = 21 °C.

The measurement of the water layer mass is carried out using the gravimetric method. The working section made of copper with a layer of liquid is placed on the scales. The wall temperature is measured with a thermocouple. Although the cell radius is constant (40 mm), the characteristic radius R_d in Equation (11) is different according to the scenario considered. In the case of water and surfactant, the length of the meniscus on the side wall of the cuvette is greater than for pure water, and the total radius R_{d1} is thus larger than for the water layer by about 11–12% at T_w = 21 °C and by 9–10% at T_w = 90 °C. Thus, to simplify the variables to a single radius when calculating the evaporation rate of the water with surfactant layer, it is necessary to divide the value of J/R_d by 1.11–1.12 and 1.09–1.1 (taking into account the length of the meniscus).

Figure 9 shows thermal images of the liquid layer and the profiles of T_s along the line 0*L*. The average surface temperature T_s of the layer increases during evaporation due to the decrease in layer height. The central region temperature is lower than at the edges. For a water layer with surfactant, the average temperature over the layer surface is higher than the temperature for a pure water layer.

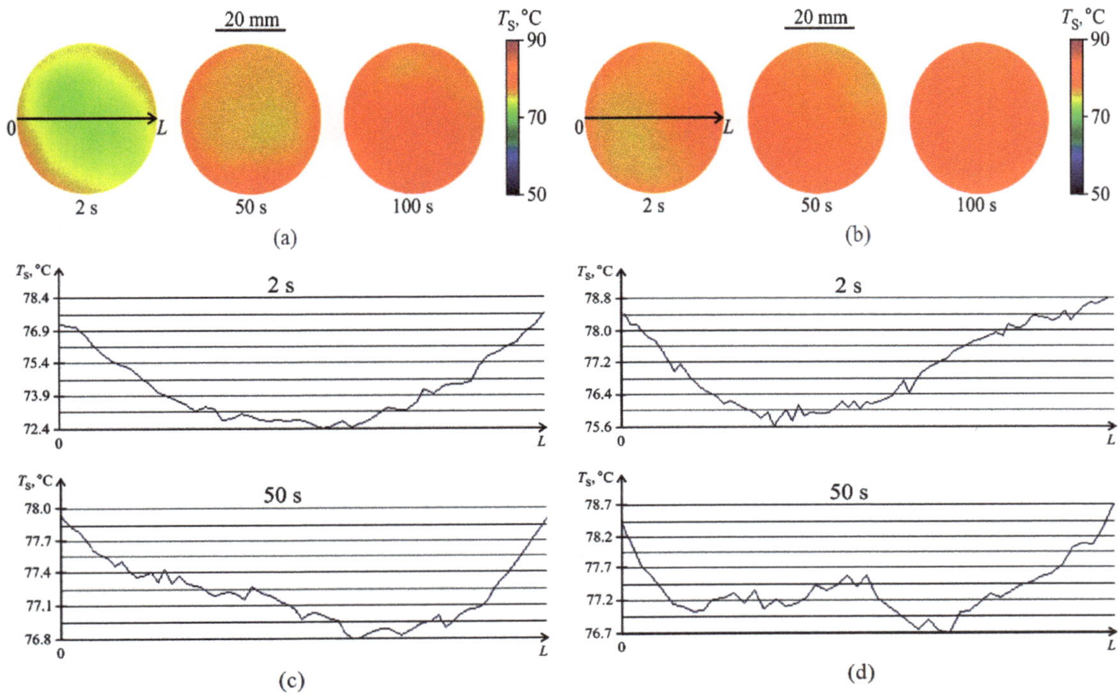

Figure 9. Thermal images of the layer interface of (**a**) pure water and (**b**) the surfactant solution. Temperature profiles of T_s along the $0L$ line for (**c**) the water layer and (**d**) the surfactant solution layer, (**a**–**d**) T_w = 90 °C, T_a = 21 °C.

4.3. Thermal Imaging Measurements of the Free Surface of a Water Droplet with and without Surfactant

A droplet of water with surfactant shows a 15–20% decrease in J_ρ (compared to pure water) at T_w = 21 °C and a 9–11% decrease at T_w = 90 °C. The obtained data correspond to the previously obtained measurements in other works. In [43], the maximum decrease in the rate of droplet evaporation with a surfactant is 25–30%. The presence of surfactant also leads to a decrease in J_ρ for the liquid layer [44]. The decrease in J_ρ is due to the competing effect of two factors—a decrease in the instantaneous area of the free surface of a drop occupied by water molecules, and a weakening of the molecular bonds between water molecules and surfactant molecules. The integral value of the two parameters for the entire free surface of the liquid decreases the equilibrium partial pressure of water vapor ρ_{vs} and, in accordance with Equation (11), the evaporation rate.

To assess the influence of nonisothermality, the temperature of the free surface of the drop T_s was measured using a thermal imager and an optical lens providing a tenfold increase in the image (Figure 10a). Evaporation leads to a decrease in T_s. The minimum drop temperature corresponds to the maximum height, i.e., the center of the drop (Figure 10b). Since the diameter of the water with surfactant drop is 1.8–2 times larger than the pure water drop diameter, the height is significantly less. As a result, the temperature difference $\Delta T_w = T_w - T_s$ is lower by 0.4–0.7 °C for a water with surfactant droplet than for water alone. In other words, a drop with a lower height has higher values of T_s and J_T. When adding a surfactant, the increase in J_T (the effect of nonisothermality) is 2–3%.

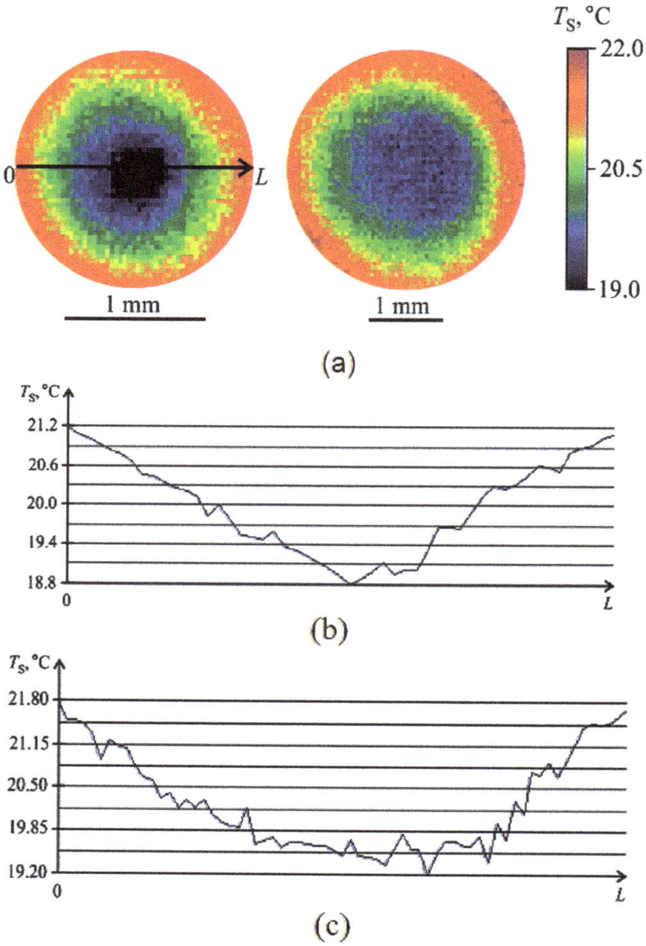

Figure 10. (**a**) Thermal images of the droplet interface for t = 20 s (without heating): water is on the left; water with surfactant (10,000 ppm) is on the right; and temperature profiles of the droplet interface at a cross section of $0L$: (**b**) water; (**c**)-water with surfactant (10,000 ppm).

With high temperature heating (T_w = 90 °C), the droplet diameter increases by 50–60% due to the addition of surfactant. Due to the fast evaporation rate on the heated wall, the temperature T_s depends much more on the drop height. For a water with surfactant droplet, the average temperature T_s (Figures 11 and 12 for t = 7 s) for the droplet surface is 4–5 °C higher than that for water during almost the entire evaporation time. This effect of nonisothermality (J_T) leads to a 14–17% increase in the evaporation rate of a droplet of water with surfactant, compared to a pure water drop (according to the water vapor equilibrium curve). Figures 11 and 12 show thermal images of the surface of a drop located on a heated wall, as well as temperature distributions of T_s along the $0L$ axis. As the wall temperature and T_s increase, the difference ΔT_s increases (comparison with Figure 10a). The small height of the drop and the high heat flux from the wall contribute to the temperature field uniformity over the drop surface.

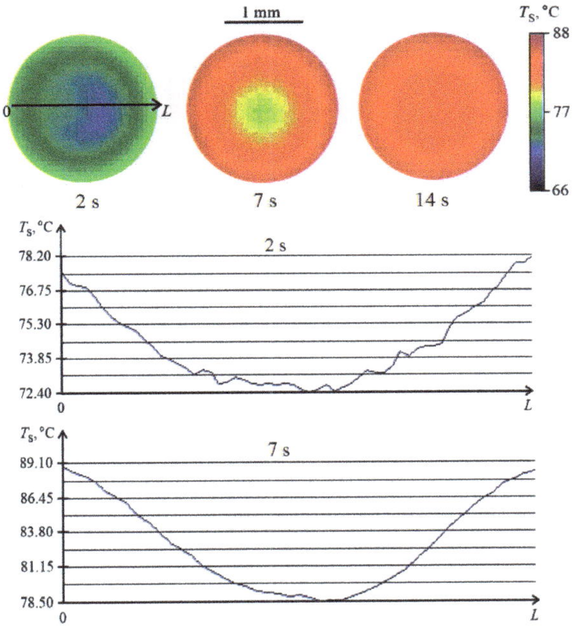

Figure 11. Thermal images of the water droplet interface and temperature profiles of the droplet interface at the cross section of 0*L* (T_w = 90 °C).

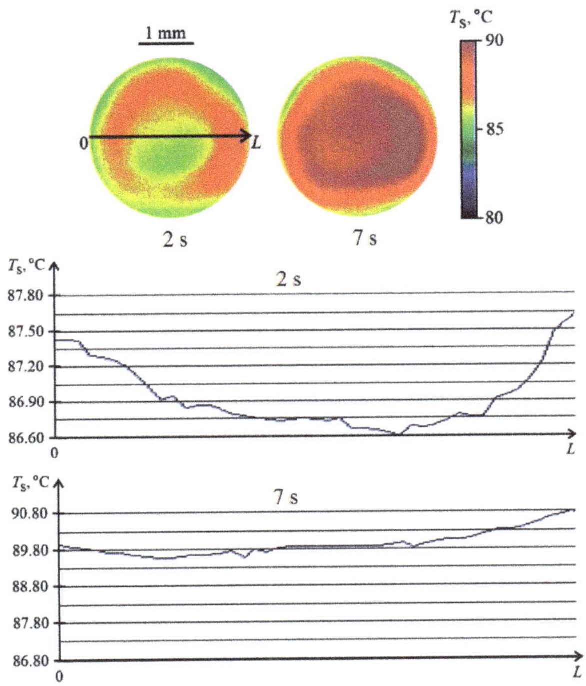

Figure 12. Thermal images of the droplet interface and temperature profiles at the cross section of 0*L* (T_w = 91 °C, water with surfactant at 10,000 ppm concentration).

The temperature profile of the droplet surface at a high wall temperature (T_w = 90 °C) is qualitatively similar to that at an unheated wall. (Figure 10b). The minimum temperature is at the center of the drop and the maximum temperature is at the edges. This distribution is associated with a negligible effect of free convection in the liquid, i.e., the temperature field is determined by the conductive heat transfer due to the small values of the Rayleigh numbers. In this case, the temperature T_s is determined by the height of the droplet ($T_s \sim 1/h$). The surfactant results in a higher temperature T_s and a more uniform temperature distribution over the droplet surface.

4.4. *Assessing the Impact of Key Factors on Evaporation of the Water Droplet with Surfactant*

To evaluate the effect of several factors on droplet evaporation, it is convenient to exclude the influence of the droplet base diameter, which strongly depends on the surfactant. For this purpose, it is convenient to attribute the value of the evaporation rate to the diameter of the droplet.

Figure 13 provides experimental data on the dependence of $J/2R_d$ on the wall temperature T_w. Points are given for pure water (points 1), water with surfactant 1000 ppm (points 2) and water with surfactant 10,000 ppm (points 3).

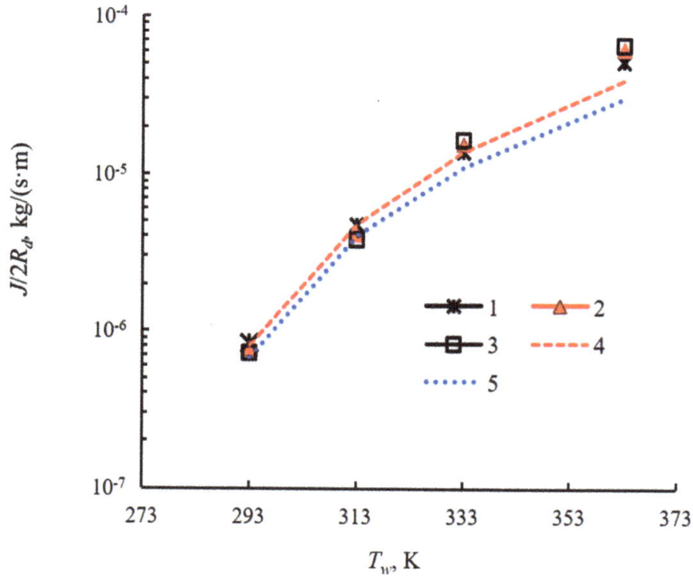

Figure 13. Drop evaporation rate depending on the wall temperature T_w. Experimental results: 1—water, 2—water with surfactant (1000 ppm), 3—water with surfactant (10,000 ppm); and by calculation from Equation (10) for the droplet contact angle: 4—of 60°, and 5— of 27°.

For droplet evaporation at room temperature, as well as at small wall heating (T_w = 313 °K), a water drop evaporates faster than a water with surfactant one (by 14–17% at T_w = 293 °K). At a temperature of T_w = 333 °K, the evaporation rates for points 1–3 are close to each other. At a wall temperature of T_w = 363 °K, a droplet of water with surfactant with a concentration of 10,000 ppm evaporates 25–27% faster. Thus, the simultaneous influence of several factors considered above leads to an inversion of $J/2R_d$ c with increasing temperature. The calculated curve 4 is constructed for the maximum value of the contact angle. This angle corresponds to the experiment with the maximum angle, which is taken as an average value over time (θ = (97° + 20°)/2 = 60°). The curve of angle change over time has a quasi-linear character. Curve 5 is plotted for the lowest contact angle over time (θ = (35° + 20°)/2 = 27°). The drop radius is also taken as the time average in the calculation and processing of the experimental data.

The diffusion coefficient of vapor in air is determined at a temperature $T = (T_s + T_a)/2$, where T_s is the temperature of the free surface of the drop, T_s is the temperature of the external air, and the temperature T_s is measured with a thermal imager. The equilibrium partial pressure of water vapor and the value of the vapor density are determined from the temperature of the free surface of the drop T_s.

Based on the above analysis, diagrams illustrating the influence of various factors on the evaporation of a drop are presented in Figure 14.

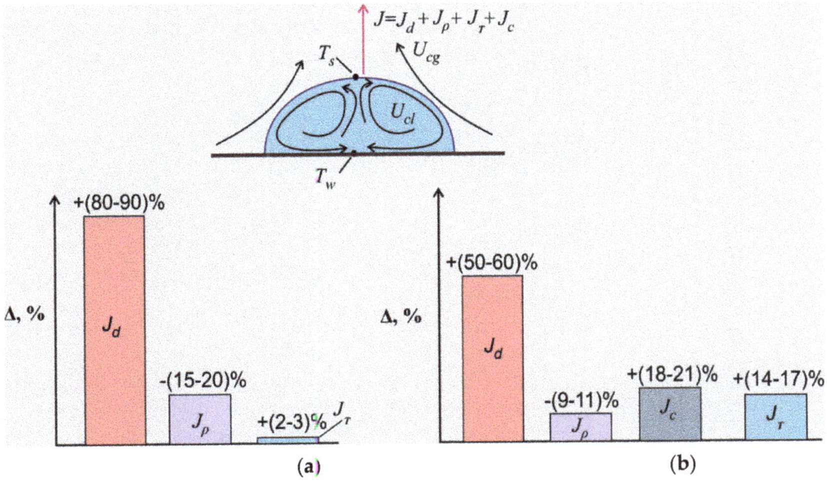

Figure 14. Influence of key factors on droplet evaporation, $\Delta = 100\% \, (J_{w,s} - J_w)/J_w$. (**a**) $T_w = 20$–21 °C, (**b**) $T_w = 90$ °C. Key: J_w (kg/s) —pure water mass flux, $J_{w,s}$ —solution mass flux for surfactant concentration of 10,000 ppm, J_d—drop diameter effect, J_ρ—partial pressure of water vapor effect, J_c—thermogravitational convection effect, and J_T—nonisothermality effect.

In Figure 14, the influence of various factors is denoted as follows: J_d is the influence of the droplet diameter, J_ρ is the influence of the equilibrium partial pressure of water vapor, J_c is the influence of thermogravitational convection, and J_T is the influence of non-isothermality. Both at $T_w = 20$ °C and at $T_w = 90$ °C, the predominant effect on evaporation is the growth of the droplet diameter due to the surfactant (J_d). At high temperature, J_c and J_T are comparable. The decrease in the evaporation rate (J_ρ) as a result of a decrease in the partial pressure of vapor at room temperature is double that at high-temperature heating. As the temperature T_s increases, the effect of J_ρ on evaporation decreases. The influence of the natural convection of gas ($J_c \sim (R_d)^{0.5}$) in Equation (5) noticeably manifests itself at high temperatures only (in Figure 10b, 18–21%) and at room temperature, thermogravitational convection can be neglected. As the wall temperature increases from 20 °C to 90 °C, the effect of J_T on the enhancement of evaporation increases approximately six times (from 2–3% to 14–17%). Thus, the $J/2R_d$ inversion with increasing wall temperature (when comparing the evaporation rate of a water drop and a drop of water with surfactant) is associated with a strong dependence of all these key factors on temperature.

At $T_w = 90$ °C, the experimental points $J/2R_d$ are located 67–71% above the calculated curves. According to Figure 14b, the influence of gas convection leads to a 67–70% excess of the experimental values of $J/2R_d$, compared to the calculated diffusion model (Equation 12), if the measured free surface temperature T_s is taken for the diffusion model. As mentioned above, 40% corresponds to an increase in the evaporation rate due to the buoyancy of water vapor. In Figure 14, this factor is not specified, since it has the same effect on a water drop with or without a surfactant. The increase in the evaporation rate due to thermogravitational convection is approximately 30% (40% + 30% = 70%). The effect of nonisothermality is expressed in a decrease in temperature T_s compared to the wall temperature T_w. Thus, for T_w = 90, 60, 40, and 21 °C, the average temperature for the droplet surface T_s is approximately 78–80 °C, 54–55 °C, 38–39 °C, and 19–19.5 °C, respectively. At the maximum wall temperature (T_w), the decrease in ρ_{vs} due to the decrease in T_s is 40–50% (the maximum effect of nonisothermality on droplet evaporation).

At T_w = 60 °C, the influence of convection and nonisothermality decreases. At T_w = 40 °C, the calculation is slightly lower than the experimental points. During evaporation without heating (T_w = 20 °C), the Rayleigh number for the gas and liquid phases is low (there is no natural convection, since there is a small droplet height and small temperature gradients due to droplet evaporation), and the experimental data are quite accurately modeled by the diffusion model.

5. Conclusions

An experimental study of the evaporation of sessile drops of aqueous solutions with surfactant under nonisothermal conditions has been carried out. The choice of an immobile liquid drop as the object of study enabled the behavior of the contact line of three phases to be investigated, and the flow of an evaporating liquid to be measured over a wide temperature range from 20 °C to 90 °C. Using the optical method, droplet contact angles, droplet heights and contact diameters of evaporating droplets were measured in the temperature range.

An analysis of the contact diameter of the evaporating droplets has shown that pure water and weakly concentrated solutions on a copper substrate are characterized by fixed contact lines (so-called pinning). However, for a highly concentrated solution (above critical micelle concentration), pinning of the droplets under study was absent. At the initial stage, when the droplet was heated, an intense increase in the droplet diameter was observed.

Surfactant presence has a complex effect on the rate of a droplet's evaporation due to several factors, some of which decrease the evaporation rate while others, vice versa, increase it:

- The largest effect on the increase in the evaporation rate of a droplet is associated with an increase in its diameter due to a decrease in surface tension.
- It is shown that the evaporation of the surfactant solution layer is slower than for a pure water layer due to the drop in the partial vapor pressure. The magnitude of the effect depends on the substrate temperature. At room temperature, the decrease in the evaporation rate is 15–20%, and for a high-temperature substrate, the change is 9–11%.
- The contribution of nonisothermality increases by a factor of six as the substrate temperature rises from 20 °C to 90 °C.
- The effect of natural gas convection is significant only for nonisothermal evaporation with a contribution of about 20%.
- The effect of free gas convection leads to an excess of the experimental values of $J/2R_d$ by 67–70% compared to the calculated diffusion model.
- The effect of surfactants on the specific evaporation rate is inverted with an increase in the substrate temperature. In the isothermal case, the surfactant solution evaporates more slowly than pure water. In the nonisothermal case, the solution evaporates faster than pure water. This is due to the different dependence of key factors on temperature.

Author Contributions: Conceptualization, S.M. and I.V.; methodology, A.S. and V.M.; software, A.S. and Y.P.; validation, S.M., V.M. and I.V.; formal analysis, S.M.; investigation, A.S., Y.P. and V.M.; resources, Y.P.; data curation, S.M. and I.V.; writing—original draft preparation, S.M. and I.V.; writing—review and editing, S.M.; visualization, V.M. and A.S.; supervision, S.M.; project administration, I.V.; funding acquisition, I.V. All authors have read and agreed to the published version of the manuscript.

Funding: This work was supported by the grants of the Russian Science Foundation, RSF 20-79-10096. The tensiometer KRÜSS K100 and KRÜSS DSA 100 were provided in accordance with the state contract of Kutateladze Institute of Thermophysics SB RAS.

Conflicts of Interest: The authors declare no conflict of interest.

Nomenclature

α	Heat transfer coefficient
β	Convective mass transfer coefficient
β_T	The thermal liquid expansion
λ	Thermal conductivity
λ_c	Capillary length
σ	Surface tension
μ	Dynamic viscosity
v	Kinematic viscosity
ρ_g	Gas density
ρ_l	Liquid density
ρ_{vs}	Vapor density on the droplet interface
ρ_a	Atmospheric density
ρ_{va}	Vapor density in the ambient atmosphere
θ	Contact angle
τ_e	Characteristic time of evaporation
Δm	The change in mass
Δt	Period of time
ΔT_s	The temperature difference between the temperature of the droplet interface and the temperature of the ambient air
ΔT_w	The temperature difference between the wall temperature and the droplet interface temperature
g	Gravitational acceleration
p_v	Partial vapor pressure
q_l	Total heat flux in the liquid
q_e	Evaporative heat flux
q_r	Radiative heat flux
q_c	Natural convection heat flux
r	Latent heat of vaporization
h	Droplet height
a	Thermal diffusivity
j	Specific mass transfer rate
J	Evaporation rate
S_d	Drop surface
V_d	Drop volume
R_d	Drop radius
F	The interface area of the liquid layer
L	The characteristic length of the diffusion vapor layer
B_M	Spalding mass transfer number
T	Temperature
T_a	Ambient atmosphere temperature
T_s	Droplet surface temperature
T_w	Substrate temperature
U_ρ	The vapor convection velocity
M_v	Vapor molar mass
U_{cl}	The average convective velocity in the droplet
U_{ca}	The air velocity due to thermogravitational convection
RH	Relative humidity
Ma	Marangoni number
Nu	Nusselt number
Ra	Rayleigh number
Re	Reynolds number
Sh	Sherwood number
Sc	Schmidt number

References

1. Smakulski, P.; Pietrowicz, S. A review of the capabilities of high heat flux removal by porous materials, microchannels and spray cooling techniques. *Appl. Therm. Eng.* **2016**, *104*, 636–646. [CrossRef]
2. Wang, J.X.; Guo, W.; Xiong, K.; Wang, S.N. Review of aerospace-oriented spray cooling technology. *Prog. Aerosp. Sci.* **2020**, *116*, 100635. [CrossRef]
3. Zhang, J.; Zhang, S.; Chunhua, W.; Xiaoming, T. Recent advances in film cooling enhancement: A review. *Chin. J. Aeronaut.* **2020**, *33*, 1119–1136. [CrossRef]
4. Khan, M.G.; Fartaj, A. A review on microchannel heat exchangers and potential applications. *Int. J. Energy Res.* **2011**, *35*, 553–582. [CrossRef]
5. Smakulski, P.; Ishimoto, J.; Pietrowicz, S. The cooling performance of the micro-solid nitrogen spray technique on the cryopreservation vitrification process: A qualitative study. *Int. J. Heat Mass Transf.* **2022**, *184*, 122253. [CrossRef]
6. Pandya, N.S.; Shah, H.; Molana, M.; Tiwari, A.K. Heat transfer enhancement with nanofluids in plate heat exchangers: A comprehensive review. *Eur. J. Mech.-B/Fluids* **2020**, *81*, 173–190. [CrossRef]
7. Gandomkar, A.; Kalan, K.; Vandadi, M.; Shafii, M.; Saidi, M. Investigation and visualization of surfactant effect on flow pattern and performance of pulsating heat pipe. *J. Therm. Anal. Calorim.* **2020**, *139*, 2099–2107. [CrossRef]
8. Deegan, R.D.; Bakajin, O.; Dupont, T.F.; Huber, G.; Nagel, S.R.; Witten, T.A. Capillary flow as the cause of ring stains from dried liquid drops. *Nature* **1997**, *389*, 827–829. [CrossRef]
9. Cahile, M.; Benichou, O.; Cazabat, A. Evaporating droplets of completely wetting liquids. *Langmuir* **2002**, *18*, 7985–7990. [CrossRef]
10. Shahidzadeh-Bonn, N.; Rafai, S.; Azouni, A.; Bonn, D. Evaporating droplets. *J. Fluid Mech.* **2006**, *549*, 307–313. [CrossRef]
11. Bonn, D.; Eggers, J.; Indekeu, J.; Meunier, J.; Rolley, E. Wetting and spreading. *Rev. Mod. Phys.* **2009**, *81*, 739. [CrossRef]
12. Maxwell, J.C. *Diffusion, Collected Scientific Papers*; Encyclopedia Britannica: Cambridge, UK, 1877.
13. Fuchs, N.A.; Pratt, J.N.; Sabersky, R.H. Evaporation and droplet growth in gaseous media. *J. Appl. Mech.* **1960**, *27*, 759–760. [CrossRef]
14. Tonini, S.; Cossali, G. An analytical model of liquid drop evaporation in gaseous environment. *Int. J. Therm. Sci.* **2012**, *57*, 45–53. [CrossRef]
15. Hu, H.; Larson, R.G. Evaporation of a sessile droplet on a substrate. *J. Phys. Chem. B* **2002**, *106*, 1334–1344. [CrossRef]
16. Gatapova, E.Y.; Semenov, A.A.; Zaitsev, D.V.; Kabov, O.A. Evaporation of a sessile water drop on a heated surface with controlled wettability. *Colloids Surf. A Physicochem. Eng. Asp.* **2014**, *441*, 776–785. [CrossRef]
17. Shekriladze, I.G. Boiling heat transfer: Mechanisms, models, correlations and the lines of further research. *Open Mech. Eng. J.* **2008**, *2*, 104–127. [CrossRef]
18. Erbil, H.Y. Control of stain geometry by drop evaporation of surfactant containing dispersions. *Adv. Colloid Interface Sci.* **2015**, *222*, 275–290. [CrossRef] [PubMed]
19. Tarasevich, Y.Y. Simple analytical model of capillary flow in an evaporating sessile drop. *Phys. Rev. E* **2005**, *71*, 027301. [CrossRef] [PubMed]
20. Ait Saada, M.; Chikh, S.; Tadrist, L. Numerical investigation of heat and mass transfer of an evaporating sessile drop on a horizontal surface. *Phys. Fluids* **2010**, *22*, 112115. [CrossRef]
21. Timm, M.L.; Dehdashti, E.; Darban, A.J.; Masoud, H. Evaporation of a sessile droplet on a slope. *Sci. Rep.* **2019**, *9*, 19803. [CrossRef]
22. Misyura, S. The dependence of drop evaporation rate and wettability on corrosion kinetics. *Colloids Surf. A Physicochem. Eng. Asp.* **2021**, *610*, 125735. [CrossRef]
23. Misyura, S.Y.; Volkov, R.S.; Filatova, A.S. Interaction of two drops at different temperatures: The role of thermocapillary convection and surfactant. *Colloids Surf. A* **2018**, *559*, 275–283. [CrossRef]
24. Misyura, S.Y. The influence of convection on heat transfer in a water layer on a heated structured wall. *Int. Commun. Heat Transf.* **2019**, *102*, 14–21. [CrossRef]
25. Truskett, V.N.; Stebe, K.J. Influence of surfactants on an evaporating drop: Fluorescence images and particle deposition patterns. *Langmuir* **2003**, *19*, 8271–8279. [CrossRef]
26. Kim, D.O.; Rokoni, A.; Kaneelil, P.; Cui, C.; Han, L.H.; Sun, Y. Role of surfactant in evaporation and deposition of bisolvent biopolymer droplets. *Langmuir* **2019**, *35*, 12773–12781. [CrossRef] [PubMed]
27. Semenov, S.; Trybala, A.; Agogo, H.; Kovalchuk, N.; Ortega, F.; Rubio, R.G.; Starov, V.M.; Velarde, M.G. Evaporation of droplets of surfactant solutions. *Langmuir* **2013**, *29*, 10028–10036. [CrossRef]
28. Guti'errez, G.; Benito, J.M.; Coca, J.; Pazos, C. Vacuum evaporation of surfactant solutions and oil-in-water emulsions. *Chem. Eng. J.* **2010**, *162*, 201–207. [CrossRef]
29. Guo, D.S.; Li, X.B.; Zhang, H.N.; Li, F.C.; Ming, P.J.; Oishi, M.; Oshima, M. Experimental study on the characteristics of temperature dependent surface/interfacial properties of a non-ionic surfactant aqueous solution at quasi-thermal equilibrium condition. *Int. J. Heat Mass Transf.* **2022**, *182*, 122003. [CrossRef]
30. Abe, Y.; Iwasaki, A.; Tanaka, K. Microgravity experiments on phase change of self-rewetting fluids. *Ann. N. Y. Acad. Sci.* **2004**, *1027*, 269–285. [CrossRef]

31. Suzuki, K.; Nakano, M.; Itoh, M. Subcooled Boiling of Aqueous Solution of Alcohol. In Proceedings of the 6th KSME-JSME Joint Conference on Thermal and Fluid Engineering Conference, Jeju City, Republic of Korea, 20–23 March 2005; pp. 21–23.
32. Savino, R.; Cecere, A.; Di Paola, R. Surface tension-driven flow in wickless heat pipes with self-rewetting fluids. *Int. J. Heat Fluid Flow* **2009**, *30*, 380–388. [CrossRef]
33. Savino, R.; Cecere, A.; Van Vaerenbergh, S.; Abe, Y.; Pizzirusso, G.; Tzevelecos, W.; Mojahed, M.; Galand, Q. Some experimental progresses in the study of self-rewetting fluids for the SELENE experiment to be carried in the Thermal Platform 1 hardware. *Acta Astronaut.* **2013**, *89*, 179–188. [CrossRef]
34. Hu, Y.; Liu, T.; Li, X.; Wang, S. Heat transfer enhancement of micro oscillating heat pipes with self-rewetting fluid. *Int. J. Heat Mass Transf.* **2014**, *70*, 496–503. [CrossRef]
35. Semenov, A.; Peschenyuk, Y.A.; Vozhakov, I. Application of Aqueous Solutions of Surfactants in Pulsating Heat Pipe. *J. Eng. Thermophys.* **2021**, *30*, 58–63. [CrossRef]
36. Sobac, B.; Brutin, D. Thermal effects of the substrate on water droplet evaporation. *Phys. Rev. E* **2012**, *86*, 021602. [CrossRef] [PubMed]
37. Kutateladze, S. *Fundamentals of Heat Transfer Theory*; Atomizdat: Moscow, Russia, 1979; p. 416.
38. Misyura, S.Y. Dependence of wettability of microtextured wall on the heat and mass transfer: Simple estimates for convection and heat transfer. *Int. J. Mech. Sci.* **2020**, *170*, 105353. [CrossRef]
39. Misyura, S.; Morozov, V.; Egorov, R. Water evaporation on structured surfaces with different wettability. *Int. J. Heat Mass Transf.* **2022**, *192*, 122843. [CrossRef]
40. Volkov, R.; Strizhak, P.; Misyura, S.; Lezhnin, S.; Morozov, V. The influence of key factors on the heat and mass transfer of a sessile droplet. *Exp. Therm. Fluid Sci.* **2018**, *99*, 59–70. [CrossRef]
41. Carrier, O.; Shahidzadeh-Bonn, N.; Zargar, R.; Aytouna, M.; Habibi, M.; Eggers, J.; Bonn, D. Evaporation of water: Evaporation rate and collective effects. *J. Fluid Mech.* **2016**, *798*, 774–786. [CrossRef]
42. Dunn, G.; Wilson, S.; Duffy, B.; David, S.; Sefiane, K. The strong influence of substrate conductivity on droplet evaporation. *J. Fluid Mech.* **2009**, *623*, 329–351. [CrossRef]
43. Doganci, M.D.; Sesli, B.U.; Erbil, H.Y. Diffusion-controlled evaporation of sodium dodecyl sulfate solution drops placed on a hydrophobic substrate. *J. Colloid Interface Sci.* **2011**, *362*, 524–531. [CrossRef]
44. Zhang, J.; Wang, B. Study on the interfacial evaporation of aqueous solution of SDS surfactant self-assembly monolayer. *Int. J. Heat Mass Transf.* **2003**, *46*, 5059–5064. [CrossRef]

energies

MDPI

Article

Effect of Monodisperse Coal Particles on the Maximum Drop Spreading after Impact on a Solid Wall

Alexander Ashikhmin, Nikita Khomutov, Roman Volkov, Maxim Piskunov * and Pavel Strizhak

Heat Mass Transfer Laboratory, School of Energy & Power Engineering, National Research Tomsk Polytechnic University, 30 Lenin Ave., 634050 Tomsk, Russia; alashikhmin24@gmail.com (A.A.); nah7@tpu.ru (N.K.); romanvolkov@tpu.ru (R.V.); pavelspa@tpu.ru (P.S.)
* Correspondence: piskunovmv@tpu.ru

Abstract: The effect of coal hydrophilic particles in water-glycerol drops on the maximum diameter of spreading along a hydrophobic solid surface is experimentally studied by analyzing the velocity of internal flows by Particle Image Velocimetry (PIV). The grinding fineness of coal particles was 45–80 μm and 120–140 μm. Their concentration was 0.06 wt.% and 1 wt.%. The impact of particle-laden drops on a solid surface occurred at Weber numbers (*We*) from 30 to 120. It revealed the interrelated influence of *We* and the concentration of coal particles on changes in the maximum absolute velocity of internal flows in a drop within the kinetic and spreading phases of the drop-wall impact. It is explored the behavior of internal convective flows in the longitudinal section of a drop parallel to the plane of the solid wall. The kinetic energy of the translational motion of coal particles in a spreading drop compensates for the energy expended by the drop on sliding friction along the wall. At *We* = 120, the inertia-driven spreading of the particle-laden drop is mainly determined by the dynamics of the deformable Taylor rim. An increase in *We* contributes to more noticeable differences in the convection velocities in spreading drops. When the drop spreading diameter rises at the maximum velocity of internal flows, a growth of the maximum spreading diameter occurs. The presence of coal particles causes a general tendency to reduce drop spreading.

Keywords: coal particle; drop impact; maximum spreading; PIV; slurry; velocity field

Citation: Ashikhmin, A.; Khomutov, N.; Volkov, R.; Piskunov, M.; Strizhak, P. Effect of Monodisperse Coal Particles on the Maximum Drop Spreading after Impact on a Solid Wall. *Energies* **2023**, *16*, 5291. https://doi.org/10.3390/en16145291

Academic Editors: Vladimir Serdyukov and Fedor Ronshin

Received: 24 June 2023
Revised: 5 July 2023
Accepted: 6 July 2023
Published: 10 July 2023

1. Introduction

In many technologies, e.g., spraying of composite liquid fuels (CLF) into combustion chambers [1–3], 3D printing, including bioprinting [4,5] and printed electronics [6], spraying of liquid friction modifiers [7], the particle-laden drops interact with various media. The development of these technological processes requires knowledge of the particle distributions in drops spreading after impact on the wall and of the effect of these particles on the flow. For example, for CLF drops, results on the effect of carbon-containing particles on drop spreading and splashing are important since the analysis of such results will help optimize the secondary atomization when hitting the wall [8], when 3D printing composite materials, tasks related to the ordering of particles on the target (functional surface) are critical [9].

In the case of particle-laden drop-wall collisions, the particle size, volume fraction, and wettability of the particles and the surface are considered as influencing factors. In this regard, the particle-laden drops can be in the form of liquid marbles [10] and slurries [11], and surfaces, respectively, can be hydrophobic [12] and hydrophilic [13]. This study focuses on the particle-laden drops interacting with a hydrophobic surface.

There is quite a wide range of works devoted to suppressing the regimes known from pure drop impact, such as jetting [14], rebound [15,16], and drop break-up during rebound [17]. It is also known that such regimes as prompt and corona splashing, which are typical for drops without particles [18–21], are significantly modified [14] since the destruction of a thin liquid layer can occur far from the drop contact line, being more similar to splashing of viscous liquids [22].

Particles in drops also affect the initial phase of interaction—spreading [11], decreasing the maximum spreading diameter D_{max} [14,15]. However, this result is rather ambiguous since there is a contradiction in which an increase in D_{max} occurs with a growth of the volume fraction of particles, but at the same time, an increase in the effective viscosity should lead to less spreading [11,23]. Obviously, in the case of particle-laden drops, spreading cannot be controlled by viscosity alone; it is a complex of influencing factors and related effects listed above. The examination of the spreading drop morphology and the distribution of internal particles results in the exploration of the spreading and receding phases [11]. In particular, Grishaev et al. demonstrated that the particle patterns strongly depend on the surface wettability, particle size, and initial impact velocity, as well as the Reynolds (*Re*) and *We* numbers [11]. These parameters also determine the spreading and receding phases.

Nicolas explored the patterns and distribution of the particles in the case of dilute slurries [23]. He observed a ring and disk-like distribution of particles as a function of *Re* and particle diameter (d_p). The manifestation of ring distribution was associated with the movement of the liquid toward the center when the drop receded. In general, the effect of particles on spreading was mainly determined by the ratio between inertial and viscous forces (*Re*). With a significant predominance of inertia ($Re > 5000$–6000), particle-laden drops spread more significantly compared to the drops without particles. However, the shape of the drop contact line was significantly distorted relative to the shape of the circle.

Since there is a rather limited amount of research on the morphology of the particle-laden drop impacting on a surface and the effect of particles on the drop spreading control, a number of issues indicated by Grishaev et al. [11] remain relevant and motivate further research. These include issues related to the influence of surface wettability, particle size, and initial drop velocities on particle distribution and drop spreading. We believe that all these issues for dilute slurries should be considered comprehensively (with empirical expressions derived), with due regard for the effective viscosity. The latter depends to a greater extent on the viscosity of the carrier medium, the buoyancy of monodisperse particles, their volume fraction, and the material. Thus, *the purpose of the study* is to experimentally determine the conditions for reducing or increasing the maximum spreading diameter of water-glycerol drops laden with hydrophilic coal particles colliding with the hydrophobic surface by analyzing the velocity fields using Particle Image Velocimetry; in addition, the research will consider the morphology of the particle-laden drop impacting on a surface—contact line deformation.

2. Materials

The solid component of the suspension was coking coal from the Berezovskaya mine, Kemerovo region, Russia. Table 1 shows the results of its elemental and technical analysis. The data was measured by the Vario microcube Elementar device. The carrier medium of a slurry was a mixture of distilled water and glycerol (Merck, Germany, CAS number: 56–81-5). The use of glycerol is caused by the need to artificially increase the viscosity of the slurry, bringing the conditions of particle movement in the flow closer to highly concentrated slurries typical of the energy sector [24,25].

Table 1. Elemental composition and main characteristics of the coking coal.

	Main Characteristics				Elemental Composition				
Substance	Fuel moisture W^a, %	Ash content of the fuel on a dry basis A^d, %	Content of volatiles V^{daf}, %	Specific heat of combustion $Q^a_{s,V}$, MJ/kg	Carbon content calculated on a dry ash-free (daf) mass C^{daf}, %	Hydrogen content calculated on a dry ash-free (daf) mass H^{daf}, %	Nitrogen content calculated on a dry ash-free (daf) mass N^{daf}, %	Sulfur content in dry matter S_t^d, %	Oxygen content calculated on a dry ash-free (daf) mass O^{daf}, %
Coking coal	2.05	14.65	27.03	29.76	79.79	4.486	1.84	0.868	13.016

The solids were kept in an air bath at 378.15 K for two hours before mixing. The dried solids were crushed in a Pulverisette 14 high-speed rotary mill with a grinding degree of 0.08–6 mm and a rotor speed of 6000–20,000 rpm. To obtain the required grinding degree, the crushed solids were sieved in an ANALYSETTE 3 SPARTAN vibrating screen with a potential sieving time of 3–20 min and using the sieves with a mesh width of 20–90 μm. The solids were weighed using Vibra AF 225DRCE analytical scales with a resolution of 10^{-6} g.

The slurries were prepared in several stages. 12 g of glycerol was added to the carrier medium—water weighing 8 g. Mixing of the aqueous solution with additives was carried out using an AIBOTE ZNCLBS-2500 magnetic stirrer with a stirring temperature of 298.15 ± 2 K, a rotation speed of 1500 rpm, and a mixing time of 7 min. Solid particles were added gradually to the mixed volume to prevent the formation of agglomerates and obtain a more uniform structure. Despite the presence of solid particles and varying their concentration, the properties of liquids within different compositions were assumed to be the same and corresponded to the water-glycerol solution under examination. This is due to the relatively low proportion of solids in the slurry. Table 2 lists the properties of the water-glycerol solution. Grinding fineness and mass concentration of coal particles varied. Two groups of coal particles by grinding fineness with a noticeable difference in size (Table 3) were chosen to reveal the additional effect of particle size on the maximum spreading diameter and the velocities in a drop. In addition, the experiments were carried out at different drop discharge heights relative to the impact surface (h) and, consequently, at different drop velocities before impact (U_0) and Weber numbers ($We = \rho D_0 U_0^2 \sigma^{-1}$, where D_0 is the diameter of the drop before impact on a wall, m; ρ is the density of the liquid, kg/m^3; σ is the surface tension of the liquid, N/m). Different Weber numbers (Table 3) separated conditionally by the four groups (We = 30; 60; 90; 120) were examined to ensure proper confidence in the potential relationships while analyzing results. The minimal one (We = 30) is defined by the experimental set-up capability, while the highest threshold (We = 120) is due to complexing the adjustment of the used optical and laser system (see Sections 3 and 4) in the case of increasing the impact droplet velocity U_0. Table 3 introduces the initial conditions for conducting experiments.

Table 2. Properties of the water-glycerol solution.

Temperature	Density, ρ	Dynamic Viscosity, μ	Surface Tension, σ
K	kg/m^3	Pa·s	N/m
293.15	1154	10.8×10^{-3}	0.06058

Table 3. Initial conditions for conducting experiments.

Discharge height—2 cm, $We \approx 30$	Coal grinding fineness	Sample name	Coal grinding fineness	Sample name	Number of experiments
	45–80 μm	-	120–140 μm	-	pcs.
Particle concentration, wt.%	0.06	Slurry 1	0.06	Slurry 3	6
	1	Slurry 2	1	Slurry 4	6
Discharge height—7 cm, $We \approx 60$	Coal grinding fineness	Sample name	Coal grinding fineness	Sample name	Number of experiments
	45–80 μm	-	120–140 μm	-	pcs.
Particle concentration, wt.%	0.06	Slurry 1	0.06	Slurry 3	6
	1	Slurry 2	1	Slurry 4	6
Discharge height—15 cm, $We \approx 90$	Coal grinding fineness	Sample name	Coal grinding fineness	Sample name	Number of experiments
	45–80 μm	-	120–140 μm	-	pcs.
Particle concentration, wt.%	0.06	Slurry 1	0.06	Slurry 3	6
	1	Slurry 2	1	Slurry 4	6
Discharge height—22 cm, $We \approx 120$	Coal grinding fineness	Sample name	Coal grinding fineness	Sample name	Number of experiments
	45–80 μm	-	120–140 μm	-	pcs.

Table 3. *Cont.*

Particle concentration, wt.%	0.06	Slurry 1	0.06	Slurry 3	6
	1	Slurry 2	1	Slurry 4	6
		Total			48

3. Experimental Set-Up

During the research, an experimental set-up was used, the scheme of which is demonstrated in Figure 1a. The PIV technique determined the flow velocity inside the drop during isothermal contact with the surface. The method estimates the velocity of convective flows in the longitudinal section of a spreading (after impact on a wall) drop parallel to the plane of the solid wall at a distance of 0.2–0.3 mm from it. The PIV method is based on recording the movement of special particles (tracers) over a very short period of time. For this purpose, "tracer" microparticles are introduced into the measured medium. The particles are illuminated by a powerful laser source, and their images are recorded by a video camera. In this case (Figure 1b), the flat laser knife generated by the radiation source cuts the drop in a plane parallel to the solid wall, and the optical axis of the video camera is perpendicular to the plane of the laser knife (Figure 1a). This method is applied due to the peculiarities of the absorption and scattering of laser radiation in the internal volume of the drop, related to the presence of solid carbon-containing particles and their mass concentration in a slurry.

The experimental set-up (Figure 1a) included the following aggregated positions: measurement (registration) part, linear displacement module, drop generation system, lighting system, high-speed video camera, and a computer for data collection and subsequent processing (not shown in Figure 1a). A platform of 0.1 m in diameter and 0.01 m in thickness with a through hole of 0.02 m in diameter was manufactured for placing an optically transparent substrate above the hole. The substrate has a diameter of 0.025 m and a thickness of 0.005 m and is made of sapphire glass (Thorlabs). The platform and the substrate with a particle-laden drop conventionally represent the measuring part.

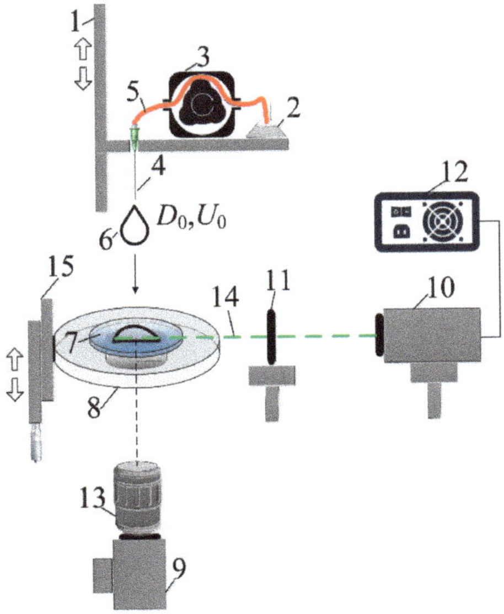

(a)

Figure 1. *Cont.*

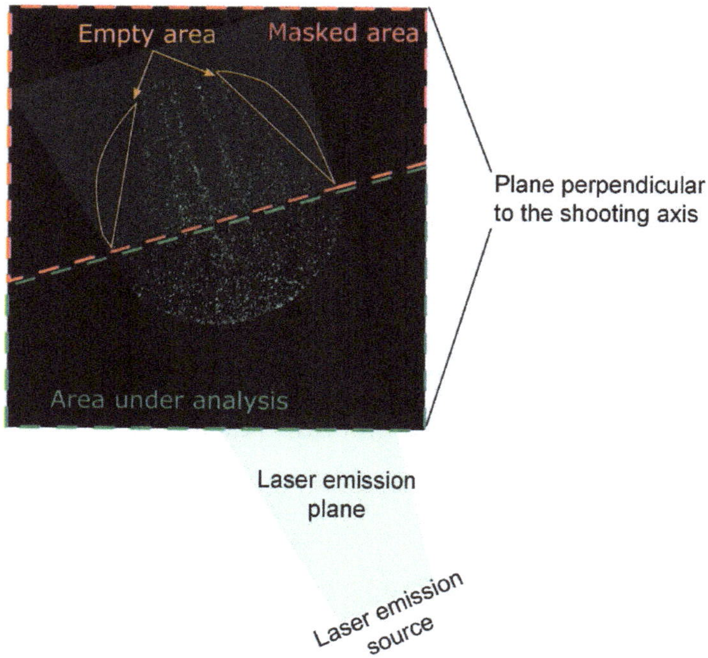

(**b**)

Figure 1. (**a**) Scheme of the experimental set-up for the implementation of the PIV method: 1—linear displacement module; 2—test tube with a liquid; 3—peristaltic pump; 4—hollow needle-nozzle; 5—silicone feed tube; 6—liquid drop; 7—optically transparent substrate; 8—a platform with a through hole for placing the substrate. 9—high-speed CMOS video camera; 10—continuous DPSS laser; 11—collimator; 12—laser power supply; 13—lens; 14—laser beam (knife); 15—micro-positioning system for the platform; (**b**) frame with the image of tracer particles in the drop under study, the planes of the laser radiation and video recording, as well as the masked and analyzed areas.

The drop discharge height varies due to the linear displacement module with an integrated servo drive and a power supply. This module is used for linear vertical (relative to the substrate surface) movement of the drop generation system, which consists of a feed tube and a hollow needle. The temperature of the liquid in the test tube corresponds to the temperature in the laboratory and is 295.15–296.15 K. The liquid in the system is pumped by a LongerPump BT100-1F dosing peristaltic pump with a set dosage volume of 0.01 mL. The drop velocity U_0 to the moment of impact on the surface could change as the drop generation system moved vertically and was 0.63 m/s, 1.17 m/s, 1.72 m/s, and 2.08 m/s. The systematic error in velocity measurement is 0.1 m/s. The diameter of the generated drop D_0 to the moment of impact on the surface remained constant and was 2.9 $\pm$ 0.05 mm. The latter was defined as the arithmetic mean of the drop diameters measured vertically and horizontally (in a frame) before the drop came into contact with the wall.

The set-up (Figure 1a) was equipped with a high-speed Phantom Miro M310 video camera recording the drop spreading and the speed of tracer particles with a sample rate of 10,000 fps and a resolution of 512 $\times$ 512 pixels. When using PIV (Figure 1a), a Nikon 200 mm $f/4$ AF-D Macro lens with a focal length of 200 mm and a relative aperture of $f/4$ was engaged. A light filter was installed on the lens, which is a laboratory orange optical glass with a bandwidth of more than 590 nm. A Thorlabs PT1B/M linear manipulator with a maximum shift of 25 mm and an accuracy of $\pm$5 μm representing a micro-positioning system allowed the adjustment of the optics relative to the platform in the vertical plane.

The minimum and maximum dimensions of the registration area were 4 × 4 mm and 25 × 25 mm, respectively. A continuous diode-pumped solid-state (DPSS) laser KLM-532A (radiation wavelength—532 nm, maximum power—5 W, power stability—3%) was employed to illuminate particle-laden drops. When implementing PIV, the generated laser radiation was transformed into a flat laser knife with an opening angle of 12° by means of the collimator based on a set of spherical and cylindrical lenses. The width and thickness of the laser knife in the measuring area were 60 mm and 0.2 mm, respectively. The plane of the laser knife was set parallel to the plane of the substrate surface. The laser knife cut the drop in the longitudinal section at a minimum distance (about 200–300 μm) from the substrate surface (Figure 1a).

Polyamide fluorescent microparticles with a diameter of 5 μm acted as tracers necessary for recording the velocity of internal convective flows in a drop. The particles absorbed laser radiation at a wavelength of 532 nm (close to the maximum of the absorption spectrum) and emitted (re-emitted) light at a wavelength of more than 550 nm. The microparticles were added to the drop at the stage of slurry preparation (see Section 2). The concentration of fluorescent microparticles was 1 g/L in accordance with the results and recommendations in Ref. [26]. The tracers were introduced into the slurry sample together with coal particles. The utilization of fluorescent microparticles, together with the optical light filter, contributed to reaching the following positive aspects. First, it made it possible to filter out reflected and refracted (by drop and substrate) laser radiation at a wavelength of 532 nm. It was possible to exclude the glare on the drop image. Second, it made it possible to filter out the light reflected from the coal particles inside the slurry drop and, as a result, to visualize only the tracers inside the drop, i.e., to determine exactly a liquid velocity.

For each liquid sample, at least three experiments were performed under identical initial conditions: drop size and velocity before the drop-wall collision. A calibration prism (Edmund Optics) with a minimum division value of 10 μm helped to determine the depth of field of the lens and the scale factor. The depth of field in the case of the PIV method was about 2 mm, and the minimum value of the scale factor for the considered cases was 0.05 mm/pixel.

4. Method of Drop Spreading Research

PIV allowed the recording of instantaneous velocity fields in a longitudinal section of the drop by fluorescent microparticle motion for an inter-frame delay. Thus, we were able to study the effect of coal particle concentration on the velocity of internal convective flows in a spreading drop. High-power laser radiation used as illumination and high-speed video recording with an exposure time of 4 μs made it possible to observe the movement of microparticles in the drop with great detail, as well as to record the microparticle velocity in spreading and receding phases. The Actual Flow v1.18 software enabled to process of experimental data by a cross-correlation algorithm when constructing instantaneous velocity fields of microparticles in a drop. The data processing included several consecutive stages (Figure 2):

- the frames recorded by the high-speed video camera were imported into Actual Flow;
- the average background intensity of the image was determined for the area without a drop; the obtained intensity was subtracted from each frame, i.e., the intensity of each pixel of the image was reduced by this value of the intensity;
- when the laser light beam is strongly distorted by a drop or when laser light is strongly absorbed in a slurry drop, a geometric mask was applied to the original frame to avoid absorption and refraction of the laser light so that only the half of the drop that the light beam enters is analyzed (Figure 1);
- each frame was divided into elementary regions of 32 × 32 pixels in size.
- for each elementary region, the correlation function was calculated, after which the coordinates of its maximum were estimated;

- the shift of the coordinates of the maximum of the correlation function in each elementary region for the time between each of two consecutive frames was determined with an accuracy of 0.2 pixels; the displacement of the coordinates of the maximum corresponds to the most probable movement of particles within the elementary region;
- using the scale factor and the time delay between two consecutive frames, the velocity of fluorescent microparticles was calculated, and plotting the corresponding velocity vector for each elementary region took place;
- using a set of velocity vectors, an instantaneous two-dimensional two-component field of the velocity of fluorescent microparticles was reconstructed at each frame;
- an interpolation procedure was performed for the obtained velocity fields, during which the modulus and direction of each velocity vector were compared with the corresponding modulus and direction of neighboring vectors, as well as vectors located in the same elementary region in the previous and subsequent velocity fields; If the differences in the modulus and direction of the velocity vector were more than 20%, the correction of these parameters for this vector occurred.

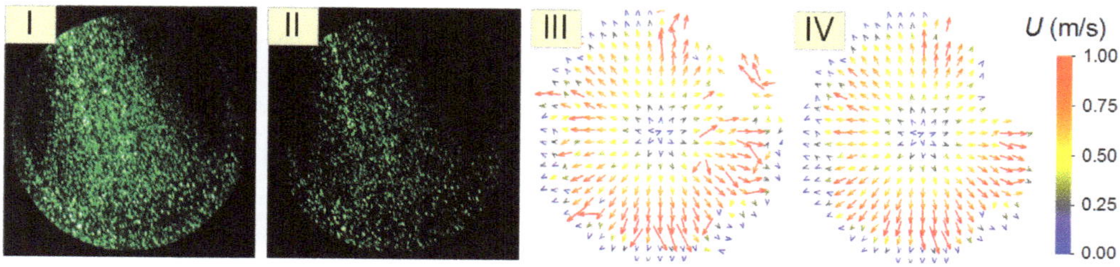

Figure 2. Illustration of the image processing and construction of convective flow velocity fields: (**I**)—primary image; (**II**)—image after subtracting the background intensity; (**III**)—instantaneous velocity field; (**IV**)—instantaneous velocity field after the vector interpolation procedure.

The following limitations and disadvantages of PIV for measuring the velocity of internal convective flows in a spreading drop were typical of preliminary experiments:

- The curvature of the drop surface causes the angle between the plane of the laser knife incident on it and the drop surface to be different from 90° [27]. At the same time, this angle also changes up during the drop spreading. This leads to the refraction of the laser knife inside the drop and, as a result, its deviation from the direction parallel to the substrate surface. In other words, the knife begins to hit the surface of the substrate and reflect off it. This fact may introduce an additional error in the measurement results;
- Due to the sphericity of the drop (curvature of its surface), the latter works as a collecting lens, focusing the laser knife falling on it [27]. This leads to the appearance of two non-laser-illuminated (shaded) "dead zones", where the convection velocity cannot be detected (Figure 1).
- When a drop spreads radially after it collides with a solid wall, waves ("horns") are often formed along its surface, the crests of which also focus the laser beam as local collecting lenses. This results in the appearance of alternating laser-illuminated and non-illuminated regions and makes the registration of convection velocities inside the drop also impossible;
- If the content of coal particles is higher than 1–2 wt.%, the laser knife does not penetrate into the drop to a depth of more than 0.2–0.5 mm due to the absorption and reflection of light by these particles. This makes it impossible to record the convection velocities inside the drop.

Thus, PIV allows measuring the convection velocity fields in a drop at any stage of its spreading over the substrate under conditions of insignificant curvature of the drop surface, even when it is heated, but only at small mass concentrations of coal particles, less than 1 wt.%. If at least one of the above disadvantages occurred, the video frame and its corresponding velocity field were not considered.

The PIV-derived results enabled analyzing the trends of maximum absolute convection velocities of internal flows in the spreading drop (U_{max}), as well as the average velocities of these flows (U_{mean}) from the contact of the surface by a drop to its maximum spreading, i.e., until the drop reaches the maximum spreading diameter D_{max}. Accordingly, the primary data are presented in the form of instantaneous maximum and average velocities of fluorescent microparticles in a drop impacting the wall (U). Instantaneous maximum velocities (U_{max}) were determined based on averaging 20 maximum velocity vectors in the frame at a given impact time (t). Instantaneous average velocities (U_{mean}) were calculated as the arithmetic mean of all velocity vectors contained in the velocity field at a specific time moment. The values of U_{max} and U_{mean} were analyzed by processing the interpolated instantaneous velocity fields of fluorescent microparticles. The data of velocity distributions from Actual Flow were additionally processed through the Wolfram Mathematica customized algorithm of a moving average filter. The results were analyzed by introducing a parameter U_{max}/U_0, representing U_{max} scaled by the initial drop velocity before the impact U_0, to describe the behavior of the maximum absolute velocities of internal flows in the spreading drop at various *We*.

Since the drop-transparent wall impact was recorded from below by means of high-speed photography, it became possible to simultaneously measure the hydrodynamic characteristics of the drop spreading, in particular, D_{max} and the time to reach D_{max}—t_{max}. The value of D_{max} was the arithmetic mean of the horizontal $D_{max}{}^{hor}$ and vertical $D_{max}{}^{vert}$ diameters since the drop spreading occurs radially and conventionally evenly from the center. To determine $D_{max}{}^{hor}$ and $D_{max}{}^{vert}$, the shadow photography method captures the collision process (Figure 3). In order to derive the empirical expressions on the spreading process, the factor of maximum spreading $\beta_{max} = D_{max}/D_0$ widely used in typical studies [18,28–30] was under examination.

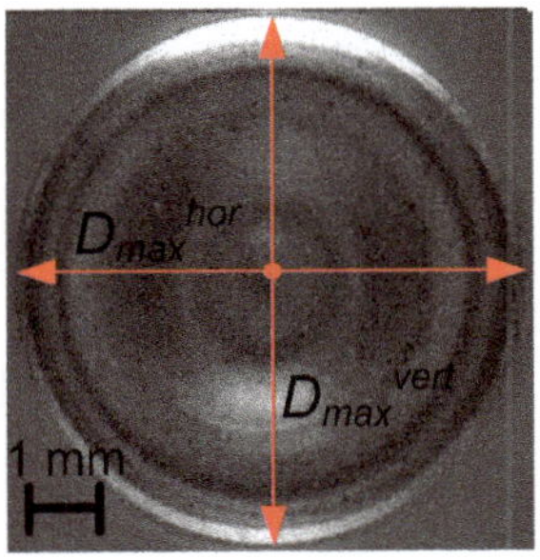

Figure 3. Illustration of the approach to analyzing the maximum drop spreading diameter D_{max} in shadow photography of the collision process.

By combining the PIV-derived data on U_{max} and the results of the drop spreading diameters recorded by shadow photography, it became possible to determine the drop diameter at which the maximum absolute velocity of internal flows D_{Umax} is reached. In addition, a parameter D_{Umax}/D_0 was introduced for quantitative and qualitative interpretation of the effect of the maximum absolute velocity of internal flows on β_{max}.

5. Results and Discussion

5.1. Morphological Observations

The results of measuring the velocity of internal fluid flow during a drop-wall collision are analyzed within four generally accepted phases (Figure 4a), such as kinetic, spreading, receding, and relaxation phases [31]. The kinetic one is characterized by a sharp increase in the maximum and average velocities of internal flows in the longitudinal section of the drop and the achievement of peak velocities U_{max} and U_{mean}. Thus, in the drop spreading phase, at the initial stage, the values of flow velocities correspond to peak values, after which they steadily tend to values close to zero. At the same time, both phases show an almost linear character of the growth and fall of U_{max} and U_{mean} values for all the considered initial experimental conditions (size and velocity of droplets, liquid viscosity). In most cases, the duration of the kinetic phase is about two times shorter relative to the spreading phase. The end of the spreading phase occurs when the drops reach D_{max}. At this point, mostly due to the viscous dissipation, the kinetic energy in the system is exhausted. The evolution of velocity fields in the longitudinal section of a spreading drop before kinetic energy depletion is presented in inserts above each velocity distribution over time (Figure 4). The inserts allow us to characterize the velocity field of internal flows as significantly inhomogeneous with a shift of increased velocities in the radial direction towards the drop spreading at the end kinetic phase and the beginning of the spreading one. During the spreading, the flow velocity decreases (see all inserts # 4 in Figure 4), starting from the periphery (rim), in the opposite direction towards the point of initial contact of the drop with the wall.

The subsequent receding phase is characterized by inertial processes, mainly in the drop rim, which leads to a local, less significant increase in the velocity of internal flows. This is clearly seen in Figure 4 from the values of U_{max} at t = 0.0075–0.0085 s. The presence of coal particles in a drop noticeably weakens the flow velocities during receding due to the effect of inhibiting the outflow of liquid by solids. This is clearly observed in the time distributions of flow velocities when comparing the ends of the receding phase in Figure 4a–e. The values of U_{max} for the compared cases are significantly different. The relaxation phase (rather, its initial stage) begins immediately after the local extremum (U_{max}) at the end of receding and is accompanied by a monotonous attenuation of internal flows.

The addition of coal particles generally affects the velocity of internal convective flows in a drop in a variety of ways. In addition, the experimental results made it possible to identify a number of influencing factors, namely, the grinding fineness of coal particles, their concentration, and *We*.

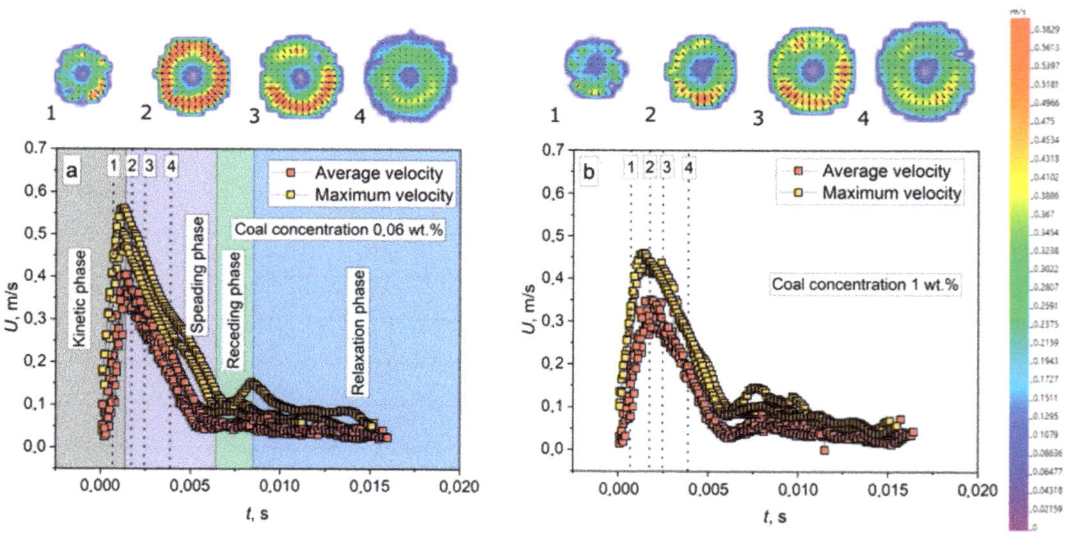

Figure 4. *Cont.*

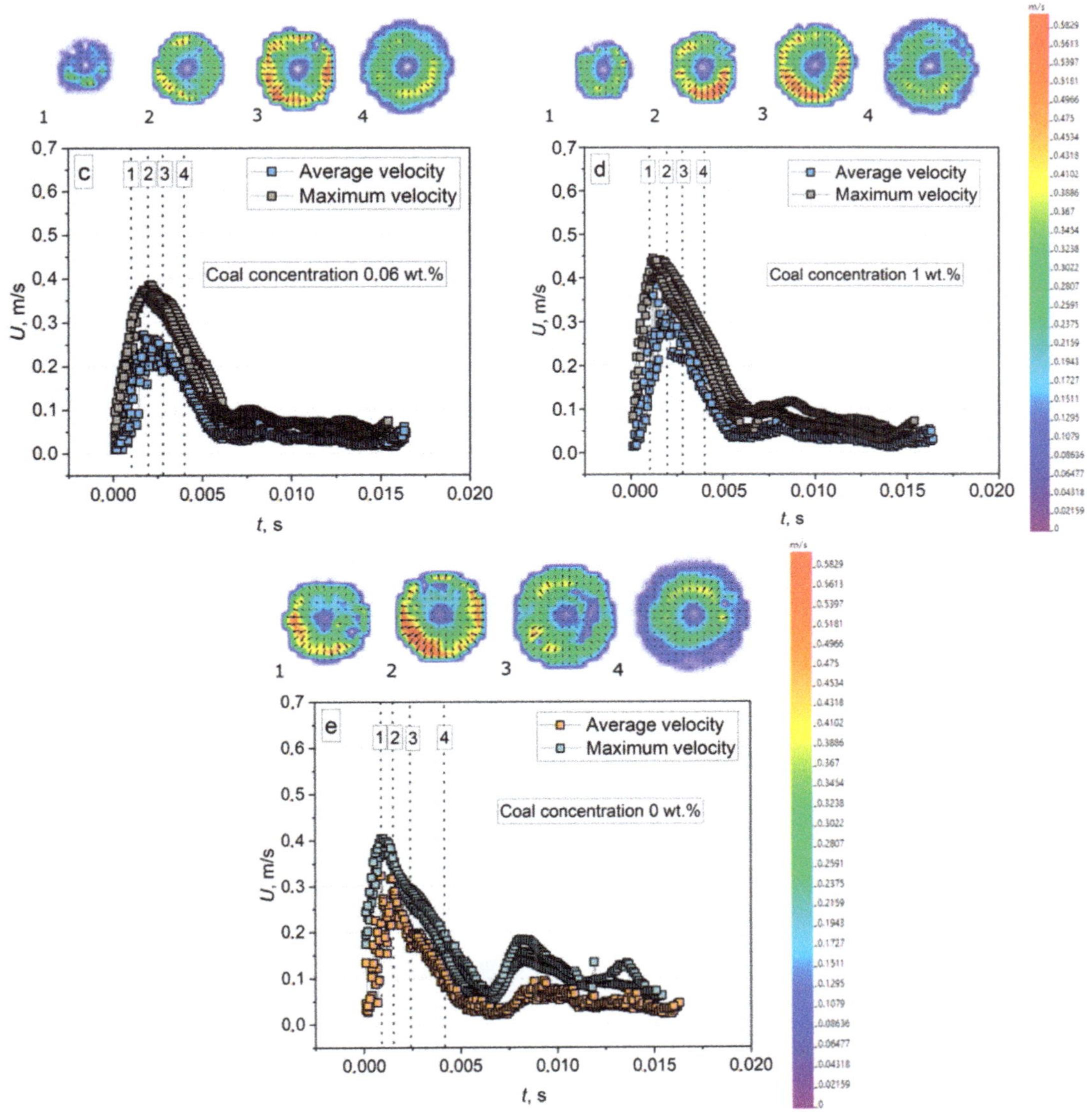

Figure 4. Instantaneous maximum and average velocities of fluorescent microparticles in a spreading drop at $We \approx 30$ as a function of the spreading time, as well as the insertion of velocity fields in the longitudinal section of the spreading drop at the following time points: 1—0.001 s; 2—0.0015 s; 3—0.0024 s; 4—0.004 s for Slurry 1 (**a**), Slurry 2 (**b**), Slurry 3 (**c**), Slurry 4 (**d**), water-glycerol solution (**e**). The confidence intervals for the experimental data were no more than 3.3%.

5.2. *Effect of We and the Concentration of Coal Particles on the Velocity of Internal Flows in a Spreading Drop*

The analysis of the results on the distributions of maximum velocities in the drop over the spreading time (Figure 5) revealed the interrelated effect of *We* and the concentration of coal particles on changes in the peak values of U_{max} in the kinetic and spreading phases. At $We = 120$, it is difficult to distinguish the effect of coal particles on the development of internal flows (Figure 5b,d), regardless of their grinding fineness and concentration. The radial motion of the drops occurs with the spreading velocity U_{spr} (Figure 5b,d) approximately twice as high as at We = 30 (Figure 5a,c). At such high values of U_{spr} (about 3.5 m/s), the inertia-driven spreading of the particle-laden drop is mainly determined by the dynamics of the Taylor rim, whose diameter becomes relatively smaller. Then, the rim begins to deform (Figure 6a,b) due to the Rayleigh-Taylor instability [32–34]. The contribution of solids is insignificant, so the values of $U_{\max}$ are quite close. While with We = 30 and a grinding fineness of 45–80 μm (Figure 5a), the addition of coal particles contributes to a rather significant increase in $U_{\max}$. However, at a coal particle concentration of 0.06 wt.%, the values of $U_{\max}$ were significantly higher than at 1 wt.%. The relative decrease in $U_{\max}$ in the case of a higher concentration can be physically associated with the formation of the internal structure (Figure 6c) and, accordingly, an increase in shear stresses between the liquid layers during the drop spreading. The latter leads to the expenditure of more energy to move the liquid. At a lower concentration, solids, having more physical space between them, cannot form the structure, and therefore the particles can act as single "accelerators" in the velocity field, which get the inertia-driven acceleration from the internal translational flow of the liquid (Figure 6d). When We = 30, and the particle grinding fineness is 120–140 μm (Figure 5c), then even in the case of a lower concentration, a certain effect of inhibition of internal flows is observed both in absolute values of $U_{\max}$ and in the spreading time t relative to the case without coal particles. This phenomenon is presumably caused by the immediate sedimentation of coal particles upon contact with the surface and their restraining disturbance of the laminar flow (Figure 6e). At 1 wt.%, the number of the particles of 120–140 μm in size becomes larger; they can not only restrain laminar flow but also mechanically deform the liquid-gas interface both on the free surface of the liquid and near the contact line after particle collisions (Figure 6f). The latter can lead to additional local liquid flows that affect the overall distribution of U over the spreading time (Figure 5c). A remarkable thing happened for the drops of Slurry 4, which characterizes the difficult-to-predict nature of particle motion. In particular, coal particles were often grouped in a rather limited area of the radially moving flow. In this case, most of the spreading particle-laden drop contained almost no coal particles.

The effect of coal particles and U_0 on $U_{\max}$ is demonstrated in the simplest format in Figure 7a. First of all, it can be seen that the values of U_0 and U_{max} are quite close, i.e., as the values of U_0 increase, $U_{\max}$ grows in direct proportion, and the momentum conservation law in a drop is satisfied. However, with more detailed observation, it is noticeable that in the absence of coal particles, the drop consumes energy by the sliding friction force when spreading along a solid surface for all We considered. In Figure 7a, this moment is expressed in the values of $U_{\max}$ for water-glycerol drops, which are lowered relative to this characteristic for particle-laden drops and the values of U_0. All other things being equal, coal particles (due to their mass) allow the development of a relatively high maximum absolute velocity of internal flows, which almost does not differ from the values of U_0. Essentially, this means that the kinetic energy (mainly) of the translational motion of coal particles in the drop compensates for the energy spent by the drop on sliding friction along the wall. Linear functions that describe the behavior of $U_{\max}$ with a change in U_0 have the following form: $U_{\max} = 1.02U_0$—0.11 for the particle-laden drops and $U_{\max} = 0.91U_0$—0.11 for the water-glycerol drops. The coefficient of determination R^2 is 0.94 and 0.99, respectively. Figure 7b, when introducing the parameter $U_{\max}/U_0$, clearly illustrates in which cases the values of $U_{\max}$ are closest to the values of U_0. In addition, Figure 7b also assumes that, based on the location of the experimental points, a further increase in We will not result in a directly proportional increase in the maximum absolute velocity of internal flows. Most likely, $U_{\max}$ will remain constant, about 2 m/s, for the considered sizes and concentrations of coal particles in the particle-laden drops. The behavior of the parameter $U_{\max}/U_0$ depending on We for water-glycerol and particle-laden drops is well described by third-order polynomial functions, Equations (1) and Equation (2), respectively. The value of R^2 for water-glycerol drops is 0.97 for the particle-laden drops—R^2 = 0.77. The value of R^2 for the particle-laden drops can be considered satisfactory since all the slurries under study are taken into account, i.e., with different concentrations and grinding fineness of coal particles.

$$\frac{U_{max}}{U_0} = 2.54 \times We^3 - 7.2 \times We^2 + 0.06 \times We - 0.77 \quad (1)$$

$$\frac{U_{max}}{U_0} = 9.3 \times We^3 - 2.78 \times We^2 + 0.03 \times We - 0.03 \quad (2)$$

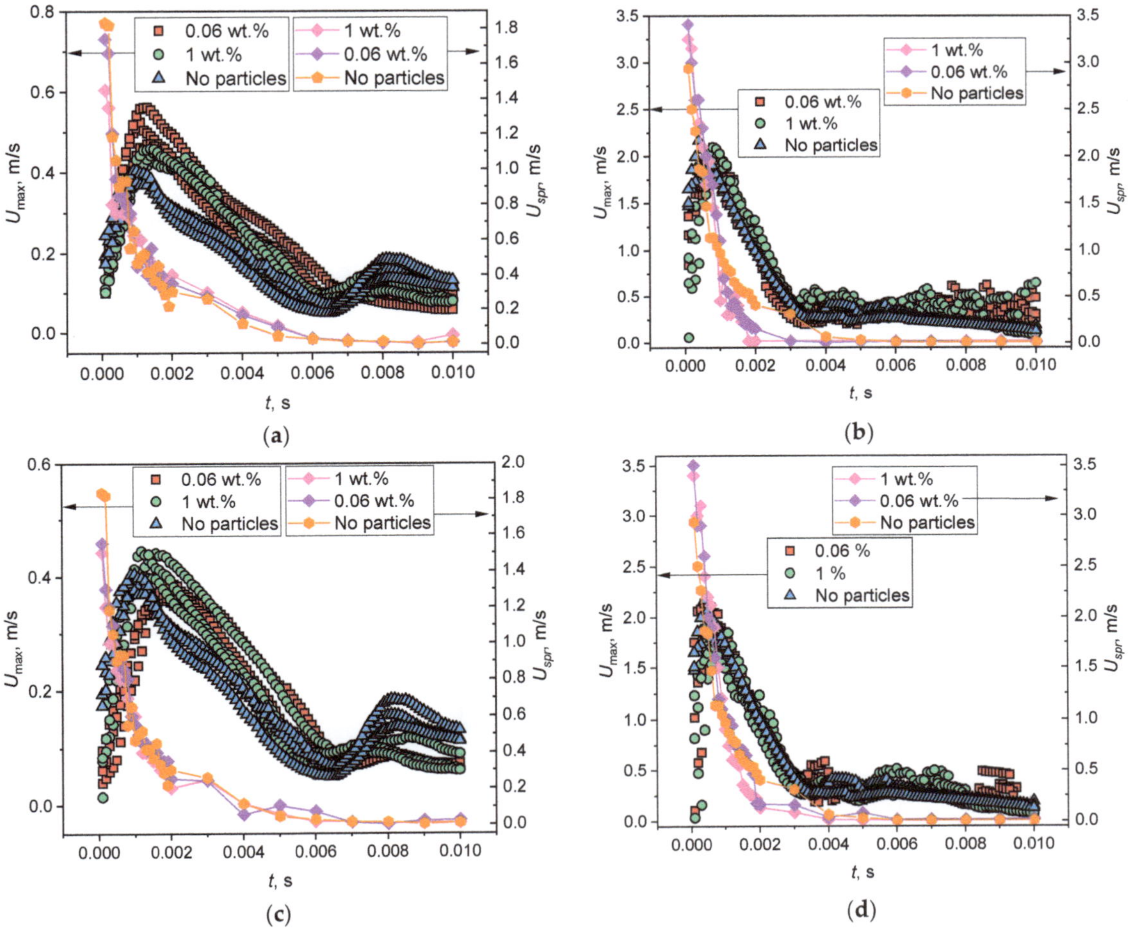

Figure 5. Effect of the Weber number and the concentration of coal particles on the distributions of the instantaneous maximum velocity of fluorescent microparticles in a drop over the spreading time: (**a**)—grinding fineness of coal particles 45–80 μm, *We* = 30; (**b**)—45–80 μm, *We* = 120; (**c**)—120–140 μm, *We* = 30; (**d**)—120–140 μm, *We* = 120. The confidence intervals for the experimental data were no more than 3.3%.

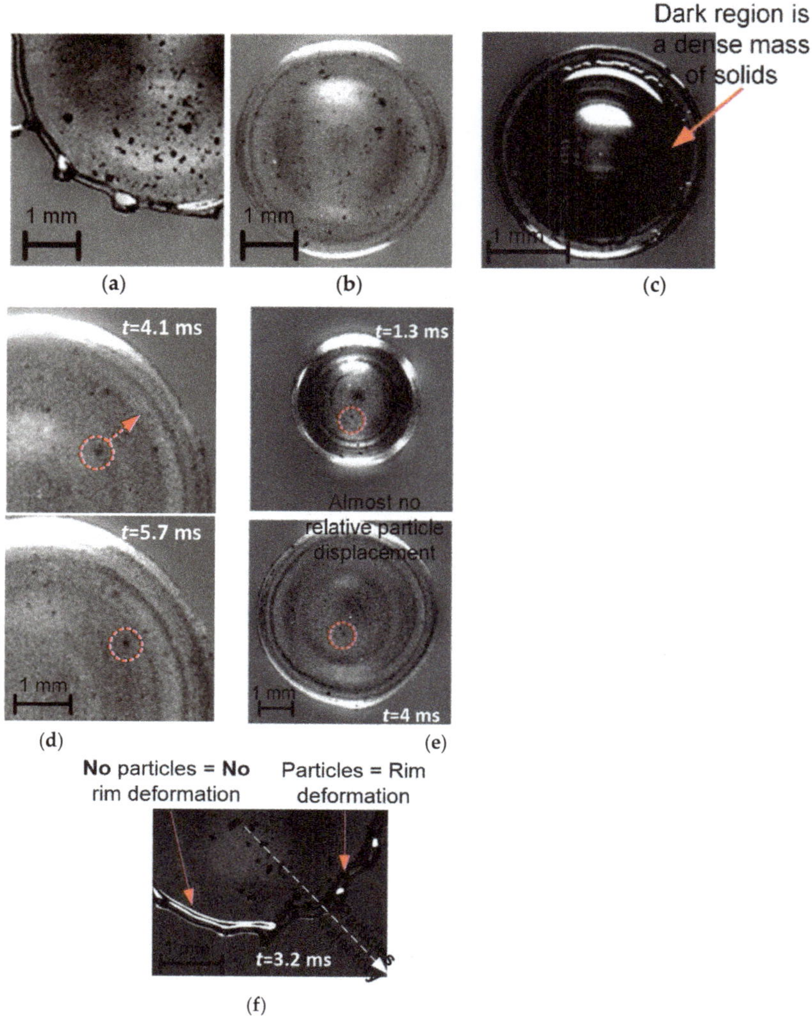

Figure 6. Frames of the forming rim during the Slurry 4 drop spreading with a particle concentration of 1 wt.% and a grinding fineness of 120–140 μm at *We* = 120 and *t* = 1.3 ms (**a**), *We* = 30 and *t* = 3 ms (**b**); frame illustrating the appearance of a Slurry 2 drop with a particle concentration of 1 wt.% and a grinding fineness of 45–80 μm at *We* = 120 and *t* = 0.6 ms (**c**); frames illustrating the relative acceleration of a single coal particle in a drop of Slurry 1 (**d**); frames illustrating coal particles deposited at the liquid–substrate interface during most of the Slurry 3 drop spreading (**e**); frame of the mechanical deformation of the Slurry 4 drop rim due to perturbation by coal particles (**f**).

5.3. *Effect of the Internal Flow Velocity on the Maximum Drop Spreading Diameter*

5.3.1. Weber Number Factor

In the previous subsection, it was shown that coal particles in a spreading drop can affect the internal flow velocities. If the velocity of internal flows in the drop increases, this should affect the spreading characteristics, in particular, the spreading diameter. Therefore, one of the key tasks of the study was to establish an implicit relationship between the velocity of internal flows in the longitudinal section of the drop and the factor of its maximum spreading β_{max}. To test this relationship, it was necessary to make sure that for all the liquids under study, the behavior of β_{max} is mainly determined by the initial drop velocity with the constancy of other terms within *We*. This confidence was achieved after summarizing the results in the framework of the relationship $\beta_{max} = f(We)$ depicted in Figure 8a. The behavior of β_{max} for particle-laden and water-glycerol drops is governed by the power function of $\beta_{max} = 0.45We^{0.4}$ with $R^2 = 0.95$.

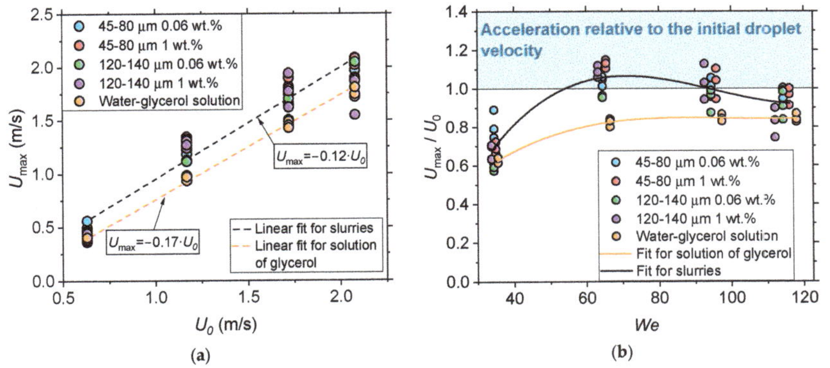

Figure 7. Maximum absolute velocities of internal flows in a drop U_{max} as a function of the initial velocity of the drop before impact U_0 (**a**), parameter U_{max}/U_0 as a function of *We* (**b**).

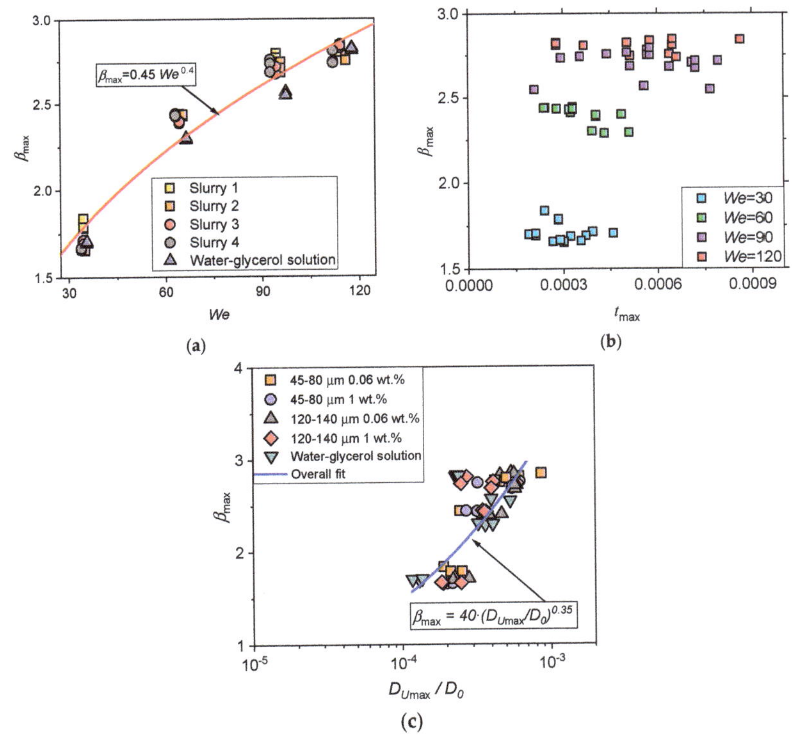

Figure 8. Maximum spreading factor β_{max} and time to reach the maximum spreading diameter t_{max} for the considered liquids at *We* = 30, 60, 90, and 120 (**a**); values of β_{max} as a function of *We* (**b**); values of β_{max} as a function of the dimensionless diameter of the spreading drop at the maximum velocity of internal flows D_{Umax}/D_0 (**c**).

Data analysis in Figure 8b suggests that the time to reach the maximum spreading diameter t_{max} for the liquids under consideration begins to vary more strongly with increasing *We*. At the same time, there are cases when t_{max} stays almost the same for different *We*. Thus, the higher *We*, the more noticeable the differences in the internal flow velocities in the spreading drops become. This is clearly demonstrated in Figure 7a. Another feature suggests that an increase in $D_{U\max}$, which depends on *We*, causes a growth of the maximum spreading diameter (Figure 8c) that is qualitatively described by a power function according to Equation (3). This result is not obvious since the later achievement of the maximum absolute velocity of internal flows in a drop is naturally in no way connected with the achievement of a higher maximum spreading diameter. Nevertheless, the established feature allows quantitative (Equation (3), $R^2 = 0.78$) and qualitative characterization of the effect of the internal flow velocity in the drop on its maximum spreading diameter.

$$\beta_{\max} = 40 \times \left(\frac{D_{U\max}}{D_0}\right)^{0.35} \quad (3)$$

Despite the same initial conditions (i.e., drop velocity before impacting the wall) for drops of all liquids within the same *We*, the experimental results in Sections 5.3.1 and 5.3.2 introduce a very ambiguous effect of coal particles on the velocity of internal convective flows in the drop. Thus, the potential energy of particles in the particle-laden drops, expressed for different slurries in the form of a change in the flow velocity when colliding with a wall, should establish a certain pattern with respect to the movement of the contact line at least until inertia-driven motion ceases (i.e., until the maximum spreading diameter is reached). This will be discussed in more detail in the next subsection.

5.3.2. Factor of Coal Particles in a Drop

The predicted pattern on the effect of the internal flow velocities $U_{\max}$ and U_{mean} on $D_{\max}$ ($D_{\max} = \beta_{\max} D_0$) was initially obtained from the point of view of the presence of coal particles in the drop, considering their grinding fineness and all the studied concentrations. In Figure 9a,b, the trend lines ($D_{\max} = aU_{\max}{}^b$ and $D_{\max} = aU_{\mathrm{mean}}{}^b$, respectively) make it clear that $D_{\max}$ for the particle-laden drops decrease relative to the water-glycerol drops at the same velocities of internal flows and drops before impacting the wall.

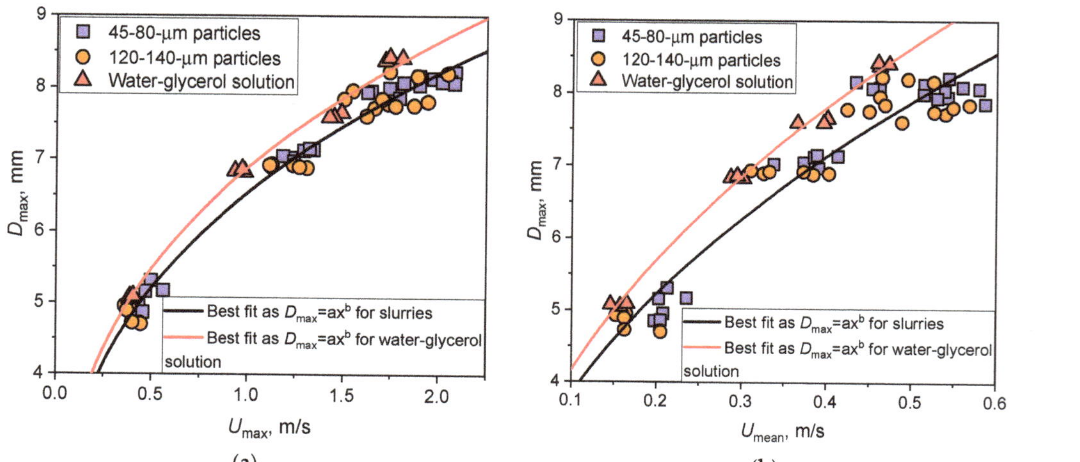

Figure 9. Effect of $U_{\max}$ (**a**) and U_{mean} (**b**) on the maximum spreading diameter at different grinding finenesses of coal particles.

The differences in $D_{\max}$ tend to grow with increasing internal flow velocities $U_{\max}$ and U_{mean}. Thus, the presence of coal particles, in general, regardless of the considered particle fineness and their concentration, causes a general tendency to decrease the intensity of liquid drop spreading.

6. Conclusions

- The PIV-derived results of measuring the velocities of internal convective flows in the longitudinal section of a water-glycerol solution drop laden with hydrophilic coal particles and spreading over a hydrophobic surface allowed the exploration of the conditions for reducing the maximum spreading diameter.
- At $We = 30$, the particle grinding fineness and their concentration strongly affect the internal flow velocities, contributing both to their increase and decrease, depending on the combination of the initial parameters of a slurry. At $We = 120$, the spreading velocity of the particle-laden drops is approximately twice as high as at $We = 30$. Given this fact, the inertia-driven spreading of the particle-laden drop is mainly determined by the dynamics of the deformable Taylor rim, and the contribution of solids is insignificant, causing the closeness of the values of the maximum absolute velocity of internal flows for various combinations of the initial parameters of a slurry. Relying on the experimental data obtained by the shadow photography and PIV, the behavior of internal convective flows in the longitudinal section of a particle-laden drop is characterized. It is revealed that the kinetic energy of the translational motion of coal particles in a drop compensates for the energy expended by the drop on sliding friction along the wall.
- The behavior of the maximum spreading factor β_{max} for particle-laden and water-glycerol drops is mainly defined by the initial drop velocity with the constancy of other terms within the Weber number and is governed by the power function of $\beta_{max} = 0.45We^{0.4}$ with a coefficient of determination $R^2 = 0.95$. Further, it is revealed the peculiarity of a noticeable increase in the differences in the velocities of internal flows in spreading drops with an increase in Weber number. Finally, as the Weber number grows, an increase in the spreading drop diameter at the maximum absolute velocity of internal flows causes the elevated values of the maximum spreading diameter and is described by an expression of $\beta_{max} = 40 \times \left(\frac{D_{U\max}}{D_0}\right)^{0.35}$. In addition, the presence of coal particles causes a general tendency to reduce the liquid drop spreading.

Author Contributions: Conceptualization, R.V., M.P. and P.S.; methodology, A.A., R.V. and M.P.; investigation, A.A., N.K., R.V. and M.P.; data curation, A.A. and M.P.; formal analysis, A.A., R.V., M.P. and P.S.; writing—original draft preparation, A.A., R.V. and M.P.; writing—review and editing, N.K., R.V. and P.S.; resources, R.V. and P.S.; project administration, M.P. and P.S. All authors have read and agreed to the published version of the manuscript.

Funding: The study was supported by the grant of the Ministry of Science and Higher Education of the Russian Federation, Agreement No 075-15-2020-806 (Contract No 13.1902.21.0014).

Data Availability Statement: The raw/processed data can be provided by the corresponding author if required.

Conflicts of Interest: The authors declare no conflict of interest.

Nomenclature

A^d	ash content of the fuel on a dry basis (%);
C^{daf}	carbon content calculated on a dry ash-free (daf) mass, (%);
D_0	initial droplet diameter before impact (m);
D_{max}	maximum spreading diameter (m);
${D_{max}}^{hor}$	maximum diameter of the drop spreading measured horizontally in a frame (m);
${D_{max}}^{vert}$	maximum diameter of the drop spreading measured vertically in a frame (m);

$D_{U\max}$	drop diameter corresponding to the maximum absolute velocity of internal flows (m);
H^{daf}	hydrogen content calculated on a dry ash-free (daf) mass, (%);
N^{daf}	nitrogen content calculated on a dry ash-free (daf) mass, (%);
O^{daf}	oxygen content calculated on a dry ash-free (daf) mass, (%);
$Q^{a}{}_{s,V}$	specific heat of combustion (MJ/kg);
Re	Reynolds number (–);
$S_t{}^{\mathrm{d}}$	sulfur content in dry matter, (%);
t	impact time (s);
$t_{\max}$	time at the maximum spreading diameter (s);
U	instantaneous maximum and average velocities of fluorescent particles in a drop impacting the wall (m/s);
U_0	initial drop velocity (m/s);
$U_{\max}$	maximum absolute velocity of internal flows in the drop during spreading (m/s);
U_{mean}	average speed of internal flows for the period from the contact of the surface with a drop and up to the maximum spreading (m/s);
U_{spr}	drop spreading velocity (m/s);
V^{daf}	amount of volatiles (%);
W^{a}	fuel moisture (%);
We	Weber number (–).
Greek symbols	
$\beta_{\max}$	maximum spreading factor (–);
ρ	density (kg/m^3);
σ	coefficient of surface tension (N/m);
μ	dynamic viscosity (Pa·s).

References

1. Wu, X.; Guo, Q.; Gong, Y.; Cheng, C.; Ding, L.; Wang, F.; Yu, G. Visualization Study on Particle Flow Behaviors during Atomization in an Impinging Entrained-Flow Gasifier. *Chem. Eng. Sci.* **2020**, *225*, 115834. [CrossRef]
2. Wu, X.; Gong, Y.; Guo, Q.; Xue, Z.; Yu, G. Experimental Study on the Atomization and Particle Evolution Characteristics in an Impinging Entrained-Flow Gasifier. *Chem. Eng. Sci.* **2019**, *207*, 542–555. [CrossRef]
3. Kuznetsov, G.V.; Strizhak, P.A.; Valiullin, T.R.; Volkov, R.S. Atomization Behavior of Composite Liquid Fuels Based on Typical Coal Processing Wastes. *Fuel Process. Technol.* **2022**, *225*, 107037. [CrossRef]
4. Zhang, Y.; Kumar, P.; Lv, S.; Xiong, D.; Zhao, H.; Cai, Z.; Zhao, X. Recent Advances in 3D Bioprinting of Vascularized Tissues. *Mater. Des.* **2021**, *199*, 109398. [CrossRef]
5. Shen, E.M.; McCloskey, K.E. Affordable, High-Resolution Bioprinting with Embedded Concentration Gradients. *Bioprinting* **2021**, *21*, e00113. [CrossRef]
6. Brian, D.; Ahmadian-Yazdi, M.-R.; Barratt, C.; Eslamian, M. Impact Dynamics and Deposition of Perovskite Droplets on PEDOT:PSS and TiO_2 Coated Glass Substrates. *Exp. Therm. Fluid Sci.* **2019**, *105*, 181–190. [CrossRef]
7. Suda, Y.; Iwasa, T.; Komine, H.; Tomeoka, M.; Nakazawa, H.; Matsumoto, K.; Nakai, T.; Tanimoto, M.; Kishimoto, Y. Development of Onboard Friction Control. *Wear* **2005**, *258*, 1109–1114. [CrossRef]
8. Thoraval, M.J.; Schubert, J.; Karpitschka, S.; Chanana, M.; Boyer, F.; Sandoval-Naval, E.; Dijksman, J.F.; Snoeijer, J.H.; Lohse, D. Nanoscopic Interactions of Colloidal Particles Can Suppress Millimetre Drop Splashing. *Soft Matter* **2021**, *17*, 5116–5121. [CrossRef]
9. Visser, C.W.; Kamperman, T.; Karbaat, L.P.; Lohse, D.; Karperien, M. In-Air Microfluidics Enables Rapid Fabrication of Emulsions, Suspensions, and 3D Modular (Bio)Materials. *Sci. Adv.* **2018**, *4*, eaao1175. [CrossRef] [PubMed]
10. Laborie, B.; Lachaussée, F.; Lorenceau, E.; Rouyer, F. How Coatings with Hydrophobic Particles May Change the Drying of Water Droplets: Incompressible Surface versus Porous Media Effects. *Soft Matter* **2013**, *9*, 4822–4830. [CrossRef]
11. Grishaev, V.; Iorio, C.S.; Dubois, F.; Amirfazli, A. Impact of Particle-Laden Drops: Particle Distribution on the Substrate. *J. Colloid Interface Sci.* **2017**, *490*, 108–118. [CrossRef] [PubMed]
12. Nguyen, T.V.; Ichiki, M. Bubble Entrapment during the Recoil of an Impacting Droplet. *Microsyst. Nanoeng.* **2020**, *6*, 36. [CrossRef] [PubMed]
13. Almohammadi, H.; Amirfazli, A. Droplet Impact: Viscosity and Wettability Effects on Splashing. *J. Colloid Interface Sci.* **2019**, *553*, 22–30. [CrossRef] [PubMed]
14. Grishaev, V.; Iorio, C.S.; Dubois, F.; Amirfazli, A. Complex Drop Impact Morphology. *Langmuir* **2015**, *31*, 9833–9844. [CrossRef] [PubMed]
15. Ok, H.; Park, H.; Carr, W.W.; Morris, J.F.; Zhu, J. Particle-Laden Drop Impacting on Solid Surfaces. *J. Dispers. Sci. Technol.* **2005**, *25*, 449–456. [CrossRef]

16. Ashikhmin, A.E.; Khomutov, N.A.; Piskunov, M.V.; Yanovsky, V.A. Secondary Atomization of a Biodiesel Micro-Emulsion Fuel Droplet Colliding with a Heated Wall. *Appl. Sci.* **2020**, *10*, 685. [CrossRef]
17. Ueda, Y.; Yokoyama, S.; Nomura, M.; Tsujino, R.; Iguchi, M. Bouncing Behaviors of Suspension Liquid Drops on a Superhydrophobic Surface. *J. Vis.* **2010**, *13*, 281–283. [CrossRef]
18. Piskunov, M.; Semyonova, A.; Khomutov, N.; Ashikhmin, A.; Yanovsky, V. Effect of Rheology and Interfacial Tension on Spreading of Emulsion Drops Impacting a Solid Surface. *Phys. Fluids* **2021**, *33*, 83309. [CrossRef]
19. Bertola, V. An Impact Regime Map for Water Drops Impacting on Heated Surfaces. *Int. J. Heat Mass Transf.* **2015**, *85*, 430–437. [CrossRef]
20. Piskunov, M.; Khomutov, N.; Semyonova, A.; Ashikhmin, A.; Misyura, S. Unsteady Convective Flow of a Preheated Water-in-Oil Emulsion Droplet Impinging on a Heated Wall. *Phys. Fluids* **2022**, *34*, 93311. [CrossRef]
21. Semyonova, A.; Khomutov, N.; Misyura, S.; Piskunov, M. Dynamic and Kinematic Characteristics of Unsteady Motion of a Water-in-Oil Emulsion Droplet in Collision with a Solid Heated Wall under Conditions of Convective Heat Transfer. *Int. Commun. Heat Mass Transf.* **2022**, *137*, 106277. [CrossRef]
22. Bolleddula, D.A.; Berchielli, A.; Aliseda, A. Impact of a Heterogeneous Liquid Droplet on a Dry Surface: Application to the Pharmaceutical Industry. *Adv. Colloid Interface Sci.* **2010**, *159*, 144–159. [CrossRef] [PubMed]
23. Nicolas, M. Spreading of a Drop of Neutrally Buoyant Suspension. *J. Fluid Mech.* **2005**, *545*, 271–280. [CrossRef]
24. Zhao, Z.; Wang, R.; Ge, L.; Wu, J.; Yin, Q.; Wang, C. Energy Utilization of Coal-Coking Wastes via Coal Slurry Preparation: The Characteristics of Slurrying, Combustion, and Pollutant Emission. *Energy* **2019**, *168*, 609–618. [CrossRef]
25. Kuznetsov, G.V.; Romanov, D.S.; Vershinina, K.Y.; Strizhak, P.A. Rheological Characteristics and Stability of Fuel Slurries Based on Coal Processing Waste, Biomass and Used Oil. *Fuel* **2021**, *302*, 121203. [CrossRef]
26. Volkov, R.S.; Strizhak, P.A. Using Planar Laser Induced Fluorescence and Micro Particle Image Velocimetry to Study the Heating of a Droplet with Different Tracers and Schemes of Attaching It on a Holder. *Int. J. Therm. Sci.* **2021**, *159*, 106603. [CrossRef]
27. Volkov, R.S.; Strizhak, P.A.; Misyura, S.Y.; Lezhnin, S.I.; Morozov, V.S. The Influence of Key Factors on the Heat and Mass Transfer of a Sessile Droplet. *Exp. Therm. Fluid Sci.* **2018**, *99*, 59–70. [CrossRef]
28. Breitenbach, J.; Roisman, I.V.; Tropea, C. From Drop Impact Physics to Spray Cooling Models: A Critical Review. *Exp. Fluids* **2018**, *59*, 55. [CrossRef]
29. Liang, G.; Mudawar, I. Review of Drop Impact on Heated Walls. *Int. J. Heat Mass Transf.* **2017**, *106*, 103–126. [CrossRef]
30. Piskunov, M.; Ashikhmin, A.; Khomutov, N.; Semyonova, A. Effects of Wall Temperature and Temperature-Dependent Viscosity on Maximum Spreading of Water-in-Oil Emulsion Droplet. *Int. J. Heat Mass Transf.* **2022**, *185*, 122442. [CrossRef]
31. Rioboo, R.; Marengo, M.; Tropea, C. Time Evolution of Liquid Drop Impact onto Solid, Dry Surfaces. *Exp. Fluids* **2002**, *33*, 112–124. [CrossRef]
32. Piskunov, M.; Breitenbach, J.; Schmidt, J.B.; Strizhak, P.; Tropea, C.; Roisman, I.V. Secondary Atomization of Water-in-Oil Emulsion Drops Impinging on a Heated Surface in the Film Boiling Regime. *Int. J. Heat Mass Transf.* **2021**, *165*, 120672. [CrossRef]
33. Burzynski, D.A.; Roisman, I.V.; Bansmer, S.E. On the Splashing of High-Speed Drops Impacting a Dry Surface. *J. Fluid Mech.* **2020**, *892*, A2. [CrossRef]
34. Sharp, D.H. An Overview of Rayleigh-Taylor Instability. *Phys. D Nonlinear Phenom.* **1984**, *12*, 3–18. [CrossRef]

energies

MDPI

Article

Nucleation of a Vapor Phase and Vapor Front Dynamics Due to Boiling-Up on a Solid Surface

Artem N. Kotov, Aleksandr L. Gurashkin, Aleksandr A. Starostin, Kirill V. Lukianov and Pavel V. Skripov *

Institute of Thermal Physics, Ural Branch, Russian Academy of Sciences, Yekaterinburg 620016, Russia; artem625@mail.com (A.N.K.); nano-studio@yandex.ru (A.L.G.); astar2006@mail.com (A.A.S.); kirill.v.lukyanov@gmail.com (K.V.L.)

* Correspondence: pavel-skripov@bk.ru

Abstract: The effect of temperature and pressure on the nucleation of the vapor phase and the velocity of the vapor front in the initial stage of activated boiling-up of *n*-pentane on the surface of a quartz fiber was studied. Using a developed approach combining the "pump-probe" and laser Doppler velocimetry methods, this velocity was tracked in the course of sequential change in the degree of superheating with respect to the liquid–vapor equilibrium line. The studied interval according to the degree of superheating was 40–100 °C (at atmospheric pressure). In order to spatiotemporally localize the process, the activation of boiling-up at the end of the light guide was applied using a short nanosecond laser pulse. A spatial locality of measurements was achieved in units of micrometers, along with a time localization at the level of nanoseconds. An increase in temperature at a given pressure was found to lead to an increase in the speed of the transition process with a coefficient of about 0.2 m/s per degree, while an increase in pressure at a given temperature leads to a decrease in the transition process speed with a coefficient of 25.8 m/s per megapascal. The advancement of the vapor front velocity measurements to sub-microsecond intervals from the first signs of boiling-up did not confirm the existence of a Rayleigh expansion stage with a constant velocity.

Keywords: pump-probe; superheated liquid; *n*-pentane; activation boiling-up; laser pulse

Citation: Kotov, A.N.; Gurashkin, A.L.; Starostin, A.A.; Lukianov, K.V.; Skripov, P.V. Nucleation of a Vapor Phase and Vapor Front Dynamics Due to Boiling-Up on a Solid Surface. *Energies* **2023**, *16*, 6966. https://doi.org/10.3390/en16196966

Academic Editors: Vladimir Serdyukov, Fedor Ronshin and Moran Wang

Received: 18 August 2023
Revised: 11 September 2023
Accepted: 3 October 2023
Published: 6 October 2023

1. Introduction

The superheating of a liquid precedes and accompanies the liquid–vapor phase transition used in various technological applications [1–6]. Interest in the issue of boiling liquids has received an additional impetus due to the development of mini- and micro-sized devices [6,7]. An important component of this problem is the stage of the vapor phase nucleation onset [8,9]. It is registered by some macroscopic response of the system, which is monitored until the superheating is removed.

The practically significant case of nucleate boiling-up on a heated surface begins with bubble nucleation in the near-wall layer of the superheated liquid. Boiling-up modes are determined by the superheat value of the liquid with respect to the liquid–vapor equilibrium line. Essentially, the initial stage of boiling represents an isothermal process of a vapor-bubble expansion, which is limited by the rate at which a mechanical impulse is transmitted to the surrounding liquid (inertial bubble growth). Since turbulence of the liquid occurs near the heating surface, the growth rate of the vapor bubble determines the hydrodynamic regime in the wall layer and the intensity of heat exchange [10,11]. In connection with the development of microelectronics, it is timely to study heat transfer from locally heated "hot spots" at the boundary with the coolant [12].

From the first experiments on the superheating of liquids, the attention of researchers was attracted to the initial stage of the boiling-up relative to the liquid–vapor equilibrium line [13]. By reducing the product $V_{liq} \cdot t_{exp}$ (where V_{liq}—volume of the superheated liquid; t_{exp}—observation time of the superheated state), it was possible to break through the background of the ready vaporization centers to achieve significant superheating. Due to the characteristic random nature of spontaneous boiling-up [14], it is difficult to study the processes in detail. The use of recording equipment is complicated by a combination of uncertainty in the waiting time for boiling-up and the relatively rapid transition to the saturation line. This is due to the difficulty of identifying the site and moment of the spontaneous boiling-up onset in a liquid at the necessary microsecond resolution for the recording of rapid processes. A significant advance in this regard was achieved with the use of low-inertia metal heaters comprised of thin wires [14,15] and films [16,17]. Due to the relatively rapid heating of wires and films, it became possible to reduce the uncertainty at the time of boiling-up. The superheating of liquid was achieved in a thin layer adjacent to the surface of the heater during non-stationary heating. Consequently, the boiling site turned out to be localized on the heating surface. Numerous experiments have been carried out with oscillography and high-speed recording of the surface boiling processes [14–24]; extensive data have been obtained on the attainable superheating of liquids during their pulsed heating [4–6]. However, there was uncertainty associated with the randomness of the boiling site on the surface and the influence of the thickness of the heated layer (the temperature gradient normal to the heating surface) on the dynamics of the observed processes. In addition, the method was mainly applied at high superheating rates to reduce the waiting time for boiling-up onset in light of the influence of surface treatment quality over longer time intervals [17,18,25].

In pursuit of further localization in terms of the location and time of observation of the initial stage of boiling-up, a technique was developed for the pulsed laser activation of the liquid–vapor transition in a miniature bubble chamber [26]. The bubble chamber [4,14] is characterized by the uniform superheating of the liquid in the capillary. Since, as a result, there is no restriction on the thickness of the heated layer, the value of the superheating temperature can be set more precisely. The development of the boiling-up process can be traced by the reflection of a probing beam following the activation pulse. This approach, when the first powerful pulse transfers the system under study to a nonequilibrium state, and the subsequent probing serves to track the relaxation process of the system, is known as the "pump-probe" method [27]. The advantage of this method lies in the strict synchronization of the processes of normalized exposure and observation, which allows for the rate of occurrence and repeatability of the observed phenomena to be assessed even at short time intervals. The purpose of the present work is to demonstrate the capabilities of the "pump-probe" laser method by identifying the features of the initial stage of activated liquid boiling-up (on the example of *n*-pentane) under the conditions of uniform superheating in a miniature bubble chamber.

2. Background

The action of the bubble chamber consists of transferring the liquid to the area of superheated states by resetting the initial pressure in the liquid p_0 below the equilibrium pressure liquid/vapor p_s (see Figure 1, A–C transition).

Thus, the degree of superheating was given by the final pressure value $p_{exp} < p_s$; the parameter monitored in the experiment was the lifetime of the superheated state prior to its decay by spontaneous boiling-up at given thermodynamic parameters. A practical basis for the detailed verification of the theory of homogeneous nucleation [8], and the measurement of different properties of substances in superheated states [4,28–30] including fuel-in-water emulsions and fuel blends [31], consisted of the determination of the temperature, pressure, and volume of the superheated liquid. To track the rapid boiling processes, a fiber optic sensor is placed in the capillary of the bubble chamber [9,29,32,33]. As an example, Figure 2 shows records of the spontaneous boiling-up signals of hexane carried out using the fiber optic sensor in previous works [9,32,33]. The change in signal voltage at the output of the photodetector is associated with a change in the density of the medium surrounding the fiber optic sensor. The letters indicating the stages of the process correspond to the states and transitions noted in Figure 1.

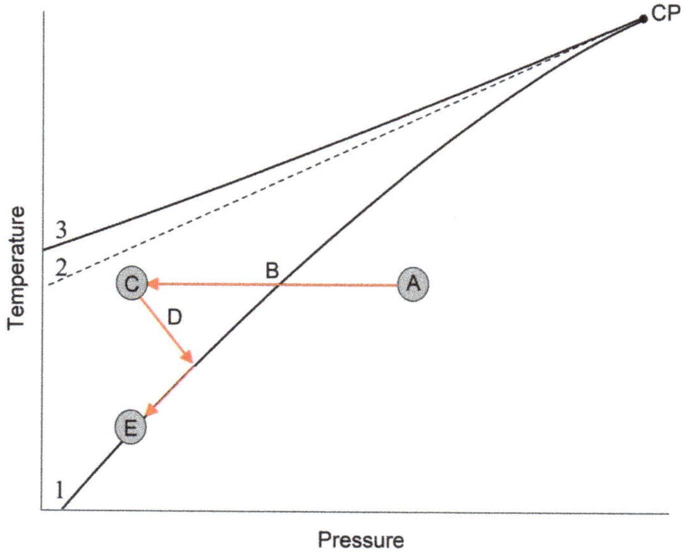

Figure 1. Liquid–vapor phase diagram designating the states of matter in the course of experiments carried out in the bubble chamber. A—initial state ($p_0 > p_s$); C—superheated state ($p_{exp} < p_s$); E—two-phase state on the saturation line; CP—critical point. Arrows B and D schematically represent the pressure relief and boiling-up development processes. 1—saturation line; 2—attainable superheat line; 3—spinodal.

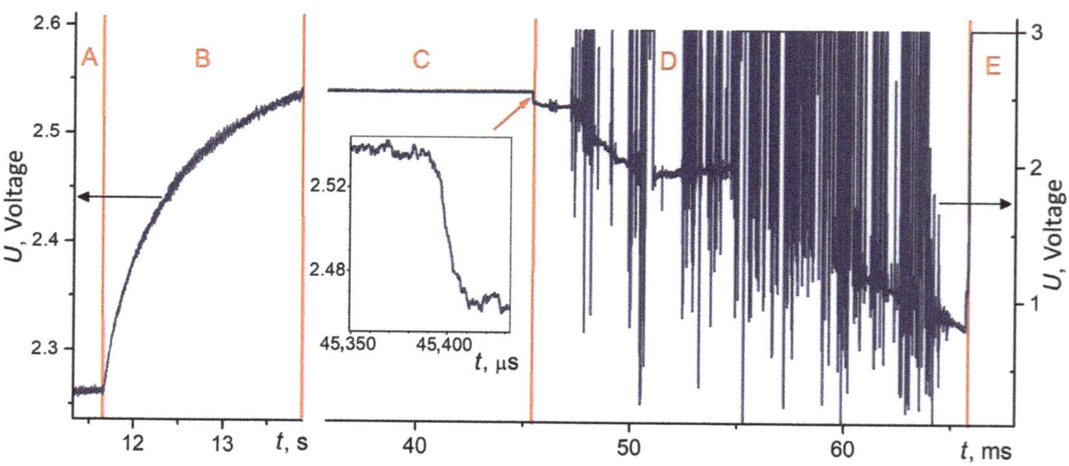

Figure 2. Voltage U of the fiber optic sensor photodetector against time in the course of experiment to study the spontaneous boiling-up of *n*-hexane. A—initial state ($p_0 > p_s$); B—depressurization; C—superheated (metastable) state ($p_{exp} < p_s$); D—boiling-up and transition to the saturation line; E—two-phase state on the saturation line.

The initial equilibrium state is indicated by A (at initial pressure $p_0 > p_s$ and temperature $T_0 = T_{exp}$). Transition B denotes the pressure release to $p_{exp} < p_s$. In state C, the liquid is superheated (metastable) until boiling-up occurs, followed by a two-phase transition, D. The process terminates in equilibrium state E on the saturation line. The described cycle can be repeated many times for a dataset to permit its statistical processing. In state C, spontaneous boiling can be expected or activated by some external action. The region of vapor phase nucleation of interest to us is located at the transition from state C to state D (marked with an arrow in Figure 2). Our approach to the boiling-up activation of *n*-pentane in a bubble chamber is described in [26]. The activation by a short laser pulse at the end of the light guide in the transparent liquid was found to have a threshold character in terms of the intensity of the excitation pulse. Increasing the energy of the pulse by changing its duration within wide limits (up to two orders of magnitude) with insufficient intensity did not lead to boiling-up. The results of a study into the effects accompanying the nanosecond excitation pulse suggest that the activation mechanism may be of a non-thermal nature. It is likely that an electrostriction mechanism of the liquid is activated at a certain intensity of the electromagnetic field of the laser pulse. Accordingly, the boiling-up activation is generated by a mechanical impulse from electrostriction. On the basis of the detected phenomenon, a method was developed for activating and studying transient the boiling-up processes across a wide area of superheated states with a precisely defined superheating temperature in the bubble chamber. The method of observing transient processes is described in [29,32]. A light guide in the bubble chamber was used to observe the liquid–vapor transition by changing the reflection signal from the end of the light guide with changes in the density of the medium [9,29,32]. Since the speed of the available photodetectors is measured in hundreds of megahertz, it is possible to record processes at nanosecond resolution.

3. Method

The application of the pump-probe method is illustrated in Figure 3. A pump pulse beam with a duration of 1–10 ns was generated by pulsed laser 1530 nm. A probe beam was continuously generated by CW laser 1550 nm. Combining the emissions for the transmission over a single fiber was achieved using the add-drop multiplexer. The pump pulse beam acted on the end of the "optofiber" on the liquid superheated in the capillary and, at a sufficient intensity, activated its boiling. The probe beam reflected from the end was returned back through the add-drop multiplexer and optical circulator to the photodetector. The intensity of the reflected signal depends on the density of the medium at the end of the optofiber and increased as the density decreased. Therefore, the boiling of the liquid was accompanied by a sharp increase in the reflected signal. In the course of boiling, a two-phase system was formed near the end of the optofiber, as shown in Figure 3. Then, the probe beam had a double reflection from the end and from the vapor–liquid interface, followed by interference. Since this interface moved as the bubble grows, the photodetector signal exhibited oscillations. The oscillation frequency is related to the speed of the interface movement according to the Doppler effect.

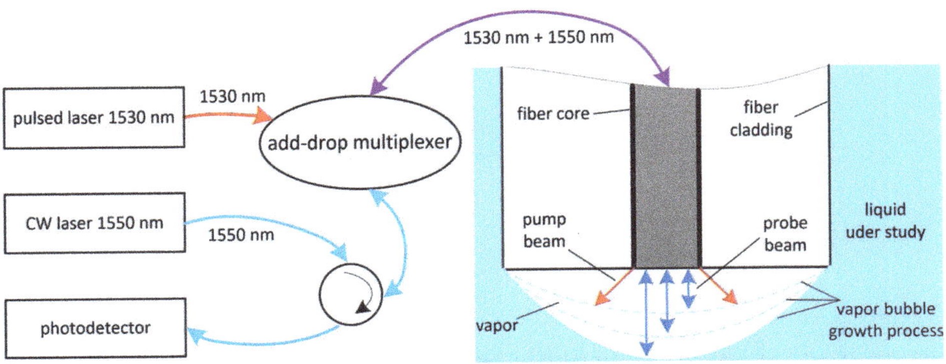

Figure 3. Scheme for the formation of optical signals by the pump-probe method for activating the boiling-up of a superheated liquid at the end of a quartz light guide.

To activate the boiling of *n*-pentane through the light guide at different temperatures and pressures, pulses of laser radiation (pump beam pulses) were applied. The required intensity and energy of the activating pulses increased with a decrease in the superheating temperature and an increase in the residual pressure p_{exp} [26]. An increase in the work of the bubble formation with a decrease in the superheating temperature and with an increase in external pressure [4,5,14] predetermined an increase in the intensity of the activating pulses. The development of the phase transition was monitored by changing the intensity of the probe beam reflected from the end of the light guide. Here, the measured intensity of the reflected radiation increased with a decrease in the refractive index of the medium according to the Fresnel formula. In turn, the refractive index decreased with a reduction in the density of the medium. The dynamics of the changes in the density of the medium were assessed by tracking the changes in the average intensity of the reflected radiation according to the photodetector signal following the activating pulse [26,34]. During the fiber-optic densitometry experiments with superheated liquid following activation of boiling-up, high-frequency oscillations were observed at the output of the high-speed photodetector of reflected radiation. However, the observed oscillations are not associated with changes in the density of the medium, appearing rather as a result of interference of the reflected rays. The traveling interference pattern shown in Figure 4, which is characteristic of the interferometers with a moving reflector, is due to the formation of an additional reflected signal as a result of the superposition of the origin and propagation of the phase interface with the reflected light waves from the stationary end of the fiber [34]. Each oscillation period in the recorded signal (Figure 4) corresponds to a shift of the vapor front by half the wavelength of the probing radiation.

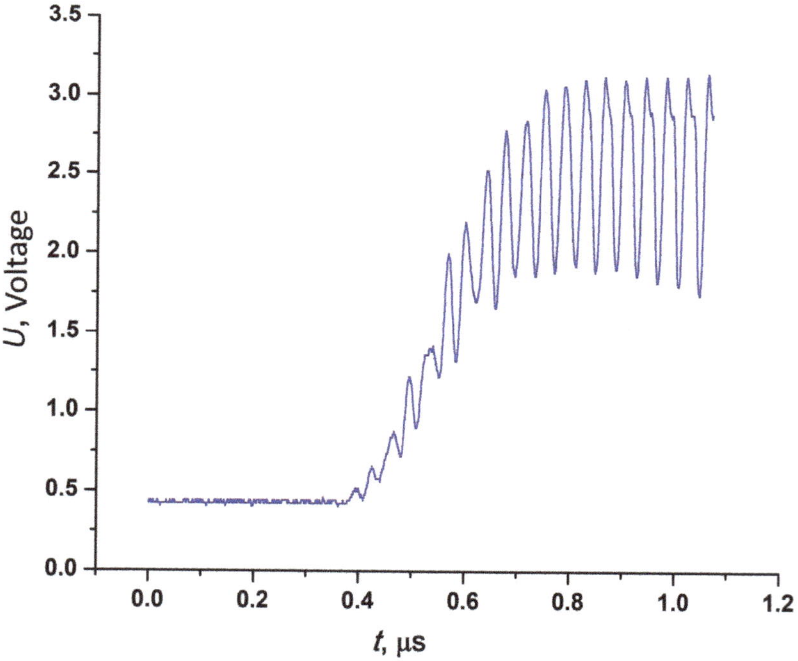

Figure 4. Structure of the reflected signal with interference of reflected rays from the end of the fiber and the moving phase interface following the activation of boiling-up of *n*-pentane at 120 °C using a laser pulse.

In this case, the frequency of the observed oscillations depended on the speed of the movement of the reflector according to the Doppler effect. Thus, the speed of movement of the reflecting phase interface can be determined by measuring the frequency of the recorded oscillations. This laser measurement method is called laser Doppler velocimetry [35]. In our experiments, the Doppler frequency shift increased in accordance with an intensification of in the bubble growth rate due to the higher superheating temperature of *n*-pentane at atmospheric pressure.

The values of the velocity of movement of the phase interface were determined by the Doppler velocimetry ratio:

$$V(t) = F_D(t) \cdot \lambda / 2n(t), \quad (1)$$

where F_D—current Doppler frequency shift (MHz); λ—wavelength of the laser diode radiation (1.55 microns); and $n(t)$—current refractive index of the medium:

$$n(t) = n_{liq} - (n_{liq} - n_{vap}) \cdot [U(t) - U_{min}] / (U_{max} - U_{min}), \quad (2)$$

where n_{liq}, n_{vap}—refractive indices of the liquid and vapor phases on the saturation line at the experimental temperature, respectively; U_{min}; U_{max}—minimum and maximum average values of the photodetector output signal corresponding to the values of n_{liq}, n_{vap}; and $U(t)$—current average value of the output signal of the photodetector.

Since the oscillations developed during decompression to the vapor phase, in most cases, $n(t)$ => 1.

4. Installation

According to the phase diagram (Figure 1), two stages are necessary to obtain a superheated state in the bubble chamber. At the first stage, the sample is heated and held in a stable liquid state (point A) at elevated pressure. At the second stage, there was a rapid decrease in the pressure (to atmospheric pressure) and the transfer of the liquid to a superheated state (point C). An experimental setup implementing such a mode is shown in Figure 5. The test liquid was used to fill a glass capillary with an internal diameter of 1 mm and a heated section of 30 mm. Inside the capillary, the liquid pressure could be set by the external pressure of nitrogen vapor on the separation membrane. To achieve this, gas was supplied from the N_2 gas tank to the gas–liquid separation block via the pressure control block. The movable membrane, which separated the nitrogen and the liquid under study while maintaining their hydraulic connection, was installed inside the gas–liquid separation block. The mobility of the membrane and the low compressibility of the liquid were necessary to ensure the equality of the pressures in the capillary and in the gas cavity of the gas–liquid separation block. Such a system allowed for two possible pressure values to be selected for the liquid in the capillary: atmospheric and that provided from the N_2 gas tank via the pressure control block. The pressure control block is controlled from a PC.

In order to heat the sample, the glass capillary was immersed in a transparent thermostat (heat carrier) with a coolant fluid (thermostat liquid). The coolant fluid was heated using an electric heater.

A precision temperature sensor was installed to monitor the temperature near the capillary. The sensor was a type K thermocouple with measurement error of 0.1 °C. To increase the accuracy of measurements, the calibration curve of thermocouples was obtained on a special bench by comparing it with the indications of a platinum thermometer. The thermostat control block maintained the set temperature of the coolant by reading the sensor indications and adjusting the electric power level of the heater. This temperature was regulated directly on the unit or via software (PC).

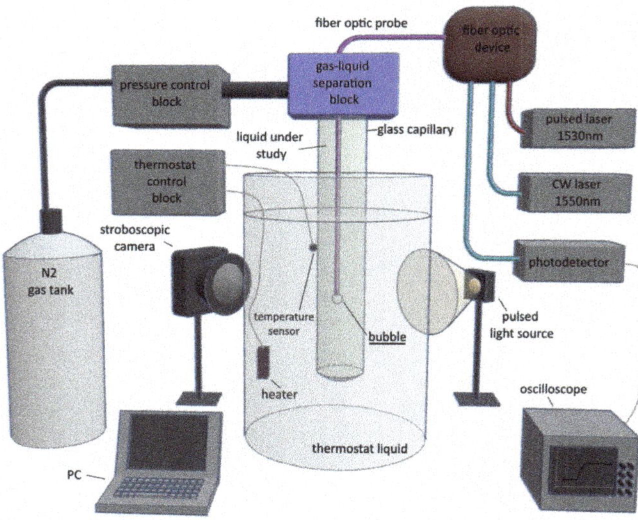

Figure 5. Block diagram of an experimental setup for the study of activated boiling-up.

A fiber optic probe was installed inside the capillary to supply the activation effect to the superheated liquid. The probe was a standard single-mode quartz fiber light guide, which was previously cleaned from the outer shell, and whose free end was located in the capillary. The outer diameter of the light guide was 125 microns; the diameter of the fiber core was 10 microns. The probe was connected to the electronic radiation supply and reception units via a fiber optic device. This scheme implemented the "pump-probe" principle. For the pump, a pulsed laser of 1530 nm, having an adjustable power of up to 30 W and a variable pulse duration on a scale from 1 ns to 100 ns, was used. The probing beam was generated by a permanent laser (CW laser 1550 nm) having a wavelength of 1550 nm and a power of 2.5 mW. Both kinds of radiation—pumping and probing—were combined and fed into the fiber probe using fiber elements comprising an adder and an optical circulator, which make up the fiber optic device. The probing radiation reflected from the free end of the light guide was separated, filtered, and then fed to the photodetector. The output of the photodetector generated an electrical signal, which was recorded using a high-speed oscilloscope. The recorded waveform files were transferred to the PC for processing. The photodetector received signals in the frequency range 0–100 MHz with a sensitivity to optical power of 0.1 V/μW. The Doppler frequency measurements using a photodetector and a Rigol 5354 oscilloscope have an uncertainty of 5% over the frequency range 10–40 MHz.

Due to the synchronous activation of the boiling-up with the pumping pulse and the subsequent growth of the vapor bubble, it was possible to capture the stages of bubble growth on a microsecond time scale using the stroboscopic video method. For this purpose, a video camera with a frame rate of 60 Hz in stroboscopic mode was used. The mode was provided by the operation of a pulsed light source, with the generation of a short (from 0.1 μs to 0.5 μs) light pulse (strobe) delayed relative to the pump pulse for a specified time. By changing the pause time between the strobe and the pump pulse, various stages of bubble growth could be recorded in a series of experiments (Figure 6).

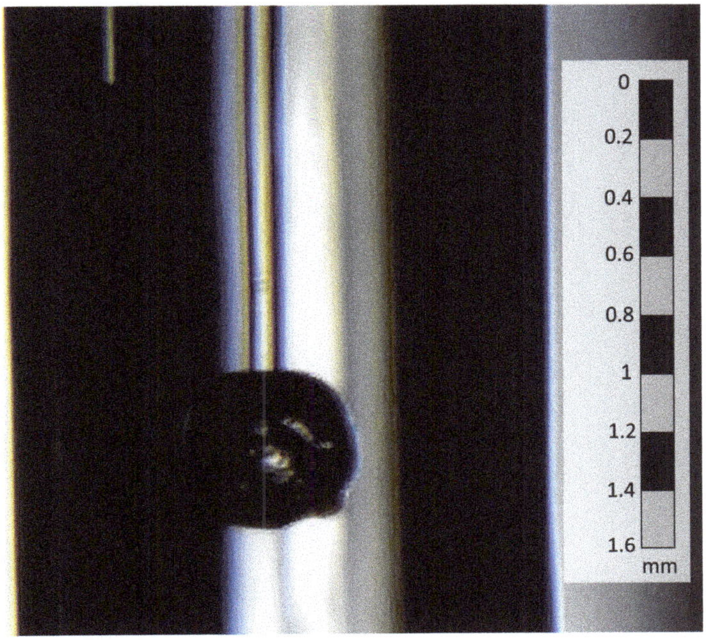

Figure 6. A vapor bubble at the end of the light guide in *n*-pentane at a delay of 50 µs from the moment of activation and an experimental temperature of 123 °C. The duration of illumination (strobe) was 0.5 µs.

5. Results

The experiments in the bubble chamber were carried out in a wide area of superheated states created by the pressure drop in a heated liquid. The mean "lifetime" of *n*-pentane before spontaneous boiling-up under the given conditions ($p_{exp} < p_s$, $T_{exp} > T_s$) ranged from units of seconds to tens of minutes. This time was sufficient to establish thermodynamic equilibrium following the pressure drop [14]. The activation of boiling was carried out by a single nanosecond pulse pump pulse (1–10 ns) with an intensity exceeding the activation threshold by ~10%. The values of the intensity and the duration of the pulse were selected experimentally from the condition of the absence of their influence on the recorded data. Figure 7 shows frames from the video footage obtained using the stroboscopic method. The duration of synchronous illumination with the pump pulse was 0.5 µs. The delay from the pump pulse increased by 1 microsecond for each subsequent frame. The video sequence shows the regular growth of the vapor film at the end of the light guide over time and the formation of a bubble after 3 µs at a temperature of 130 °C and at atmospheric pressure. Despite the image distortion due to light refraction, it is possible to estimate the size of the vapor cavity relative to the size of the light guide (125 µm) after 3 µs at a value of about 60 µm (average growth rate is 20 m/s). It is noteworthy that, for a short period of time, the surface tension forces hold the bubble at the end of the fiber. The subsequent development of the process was less predictable with the increasing influence of hydrodynamic disturbances.

An example of the recorded dependences of the voltage in the photodetector on time is shown in Figure 8. The waveforms of signals for *n*-pentane temperature values of 80 °C, 95 °C, and 130 °C are shown (superheating $T_{exp} - T_s$ is 43, 58, and 93 degrees, respectively).

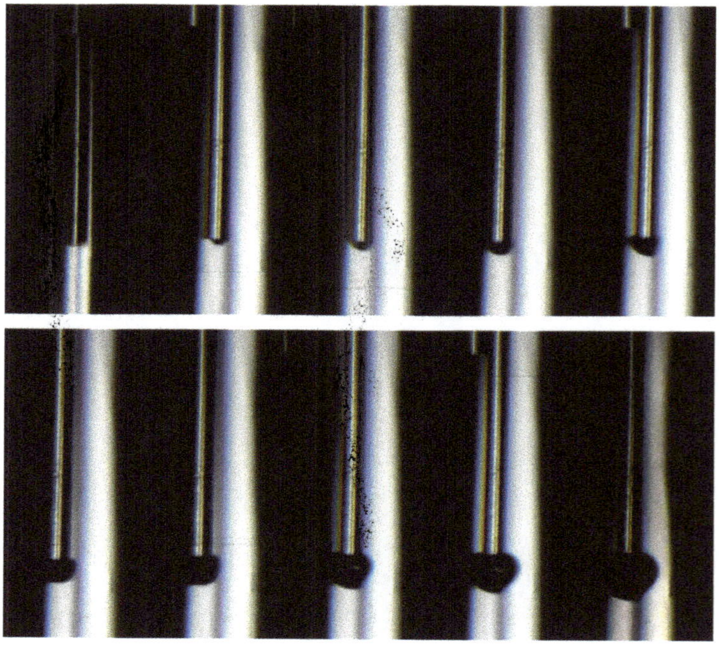

Figure 7. Bubble growth in *n*-pentane at the end of the light guide after the pump pulse. The images were obtained by stroboscopic shooting with an increase in the pulse illumination delay by 1 microsecond for each subsequent image. The experiment was carried out at a temperature of 130 °C and atmospheric pressure. The duration of illumination (strobe) was 0.5 μs.

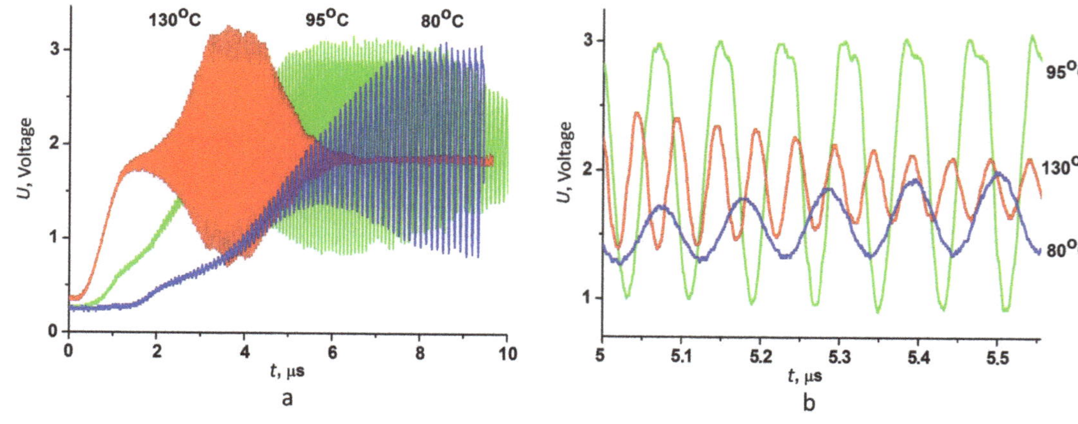

Figure 8. Measured reflection signals from the end of the light guide following activation of *n*-pentane boiling at different temperatures and atmospheric pressure on a general (**a**) and enlarged (**b**) scale with different frequency of Doppler shift at an activated phase transition with different conversion rates.

A change in the dynamics of signal growth following the activation of the boiling-up is shown in Figure 8a, while Figure 8b depicts the change in the frequency of the Doppler shift depending on temperature. The transition process to the steady-state average value occurred over a time from 1 μs for 130 °C to 6 μs for 80 °C. The frequency of Doppler oscillations varied accordingly from 12 to 25 MHz. The change in the oscillation frequency of up to 10% observed at intervals of estimating the rate of processes was associated with a known decrease in the bubble growth rate over time [11]. For the subsequent calculations, averaged values were used.

Figure 9 shows the dependences calculated from the frequency of oscillations of the vapor front velocity on time at different temperatures T_{exp}. Since a determination of the frequency of oscillations and velocity was possible only when a sufficient amplitude of the signal was reached, there is a delay in the beginning of measurements in Figure 9.

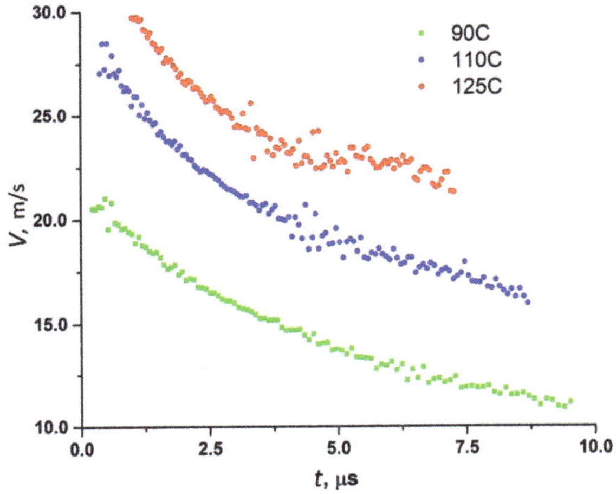

Figure 9. Change in the velocity of the vapor front over time at different initial temperatures of superheated n-pentane. The sampling delay increases for lower temperatures due to a decrease in the amplitude of the recorded signal at the beginning of the process.

The developed Doppler velocimetry technique was validated for measuring the velocity of the evaporation front on samples of pure n-pentane at various degrees of superheating relative to the liquid–vapor equilibrium temperature when the pressure drops to atmospheric values was carried out (Figure 10a). In particular, for a temperature of 130 °C, the average growth rate of 19.2 m/s agrees quite well with the photographic estimate of 20 m/s (see Figure 7). Figure 10b shows the effect of the final pressure p_{exp} in the course of its release at a given temperature T_{exp}.

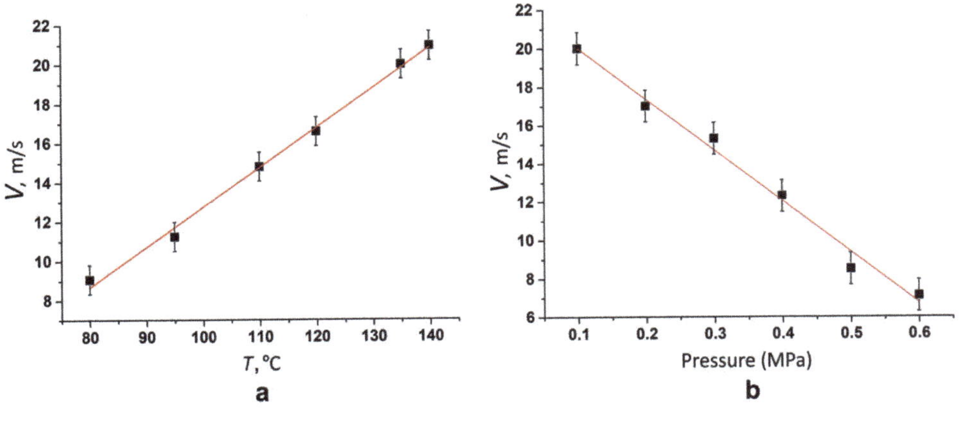

Figure 10. Temperature and baric dependencies of the initial velocity of the phase boundary displacement calculated by the Doppler velocimetry method at atmospheric pressure (**a**) and at a temperature of 135 °C (**b**).

6. Discussion

Boiling-up is commonly defined as the process of separation of the liquid and vapor phases inside a superheated liquid. For our case of relatively high superheating (40–100 °C for pentane), the formation of a vapor film on the surface of the fiber end face was observed (see Figure 7). The dynamics of the appearance and growth of a vapor film at the boundary of a solid surface and a highly superheated liquid significantly determine the intensity of heat transfer at the liquid boiling-up onset [36]. The approach combining pump-probe and laser Doppler velocimetry made it possible to track the vapor front velocity at the initial stage of *n*-pentane boiling in a miniature bubble chamber with a successive change in the degree of superheat.

The first obtained result consists of the possibility of such a local impact occurring in which the subsequent boiling process proceeds independently of the pump pulse parameters. In this case, a fairly accurate replication of the video frames and waveforms is capable of being repeatedly reproduced. The values of the initial growth rate of the vapor phase at the end of the fiber calculated from the measured Doppler frequencies differ significantly (by about 2 times) from those calculated using the Rayleigh formula for the inertial stage. Rayleigh's formula for inertial bubble growth [11] assumes the constancy of the growth rate $\dot{R}$ for bubbles with a radius R of more than 1 μm in a superheated under the conditions of our experiments liquid:

$$\dot{R} = \sqrt{\frac{2\Delta p}{3\rho}} \quad (3)$$

where Δp is the difference between the pressure in the bubble and the pressure in the liquid, and ρ is the density of the liquid.

Despite the advance of measurements of the vapor front velocity to sub-microsecond intervals from the beginning of boiling-up, we were not able to identify a region of expansion with a constant velocity as predicted by Rayleigh's formula. Figure 8a shows a gradual increase in the signal with time at the output of the fiber-optic densitometer, which corresponds to the gradual density decrease in the medium near the end of the fiber. The dependences of the vapor cavity growth rate on time and temperature turned out to be more consistent with thermally controlled bubble growth (see Figures 9 and 10). Despite the process occurring near the hot surface of the end of the fiber, the obtained results confirm the assumptions made earlier [37] about a lower vapor pressure and the probable cooling of the vapor film already at the initial stage of growth. The obtained dependencies of the initial velocity of the phase boundary displacement on temperature and pressure are close to linear (see Figure 10). The opposite effects of the experimental temperature and pressure are consistent with the change in the degree of superheating of the liquid (see Figure 1). An increase in temperature increased the speed of the boundary displacement by a factor of about 0.2 m/s per degree, while an increase in pressure reduced this speed by a factor of ~25.8 m/s per megapascal.

7. Conclusions

A method for measuring the velocity of the vapor front in the course of the activation of liquid boiling-up at the end of a light guide due to a short laser pulse has been developed and tested. The experiments were carried out in a miniature bubble chamber of the type traditionally used to determine the dependencies of the mean "lifetime" of a liquid and rate of nucleation on the temperature of superheating under strictly controlled conditions. The main novelty of our approach consists of the achievement of spatial localization in the units of micrometers with localization in time at the level of units of nanoseconds. The combination of a fiber-optic sensor with a bubble chamber according to the pump-probe method has opened up new opportunities for studying the fast-flowing processes during initial stages of liquid–vapor phase transitions using densitometric and velocimetric techniques. The obtained results, which do not fit into the Rayleigh scheme, can be interpreted as confirming the assumption of a lower vapor pressure and cooling of the vapor film already at the early stage of vapor phase growth. By applying this approach to the study of the initial velocity of the vapor front, it becomes possible to test theoretical ideas about the development of a phase transition at the interface between a liquid and a solid surface depending on the degree of superheating of the liquid.

Activation by nanosecond pulses was used to involve a wide area of moderately superheated states in the study, which were previously considered difficult to access due to the long waiting times for spontaneous boiling-up. Further research will be carried out in the field of relatively low superheating, which has significance for various practical applications, such as alternative fuel technologies based on rapid secondary atomization of composite droplets and modern minimally invasive medical tools, see references [31,34] and bibliography therein.

Author Contributions: Conceptualization, A.A.S. and P.V.S.; methodology, A.N.K. and A.A.S.; software K.V.L.; hardware A.N.K. and A.L.G.; validation, A.N.K., A.A.S. and P.V.S.; writing—original draft preparation, A.N.K., A.A.S., A.L.G. and P.V.S.; writing—review and editing, A.N.K., A.A.S. and P.V.S. All authors have read and agreed to the published version of the manuscript.

Funding: The investigation has been conducted at the expense of a grant of the Russian Science Foundation (project № 19-19-00115-P).

Data Availability Statement: Data are available on request due to organization rules.

Conflicts of Interest: The authors declare no conflict of interest.

Abbreviations

	Definition
Subscripts	
D	Doppler
exp	Experimental
liq	Liquid
max	Maximum
min	Minimum
s	Saturation
vap	Vapour
Variables and functions	
Δp	Difference between the pressure in the bubble and in the liquid
λ	Wavelength of the probing radiation
ρ	Density of the liquid
$F_D(t)$	Doppler frequency shift
$n(t)$	Current refractive index of the medium
n_{liq}	Refractive index of the liquid phases
n_{vap}	Refractive index of the vapor phases
p_0	Initial pressure in the liquid
p_s	Equilibrium pressure liquid/vapor
$\dot{R}$	Bubbles growth rate
T_0	Initial liquid temperature
T_{exp}	Experimental temperature
T_s	Equilibrium temperature liquid/vapor
t_{exp}	Observation time of the superheated state
U	Voltage of the fiber optic sensor photodetector
$U(t)$	Current value of the output signal of the photodetector
$U_{\max}$	Maximum average values of the photodetector output signal
$U_{\min}$	Minimum average values of the photodetector output signal
V_{liq}	Volume of the superheated liquid
$V(t)$	Velocity of movement of the phase interface

Abbreviations

CW	Continuous wave
CP	Critical point
States	
A	Initial state ($p_0 > p_s$)
B	Depressurization
C	Superheated state ($p_{exp} < p_s$)
D	Boiling-up and transition to the saturation line
E	Two-phase state on the saturation line

References

1. Berthoud, G. Vapor Explosions. *Annu. Rev. Fluid Mech.* **2000**, *32*, 573–611. [CrossRef]
2. Sazhin, S.S. *Droplets and Sprays: Simple Models of Complex Processes*; Springer: Cham, Switzerland, 2022.
3. Antonov, D.V.; Fedorenko, R.M.; Yanovskiy, L.S.; Strizhak, P.A. Physical and Mathematical Models of Micro-Explosions: Achievements and Directions of Improvement. *Energies* **2023**, *16*, 6034. [CrossRef]
4. Skripov, V.P.; Sinitsyn, E.N.; Pavlov, P.A.; Ermakov, G.V.; Muratov, G.N.; Bulanov, N.V.; Baidakov, V.G. *Thermophysical Properties of Liquids in the Metastable (Superheated) State*; Gordon and Breach Science Publishers: London, UK, 1988.
5. Debenedetti, P.G. *Metastable Liquids: Concepts and Principles*; Princeton University Press: Princeton, NJ, USA, 1996. [CrossRef]
6. Ching, E.J.; Avedisian, C.T.; Cavicchi, R.C.; Chung, D.H.; Rah, K.J.; Carrier, M.J. Rapid Evaporation at the Superheat Limit of Methanol, Ethanol, Butanol and n-Heptane on Platinum Films Supported by Low-Stress SiN Membranes. *Int. J. Heat Mass Transf.* **2016**, *101*, 707–718. [CrossRef] [PubMed]
7. Khan, S.; Atieh, M.; Koç, M. Micro-Nano Scale Surface Coating for Nucleate Boiling Heat Transfer: A Critical Review. *Energies* **2018**, *11*, 3189. [CrossRef]
8. Ermakov, G.V.; Lipnyagov, E.V.; Perminov, S.A. Classical Theory of Homogeneous Nucleation in Superheated Liquids and Its Experimental Verification. *Thermophys. Aeromechanics* **2012**, *19*, 667–678. [CrossRef]
9. Gurashkin, A.L.; Starostin, A.A.; Ermakov, G.V.; Skripov, P.V. Communication: High Speed Optical Investigations of a Character of Boiling-up Onset. *J. Chem. Phys.* **2012**, *136*, 021102. [CrossRef] [PubMed]
10. Pavlenko, A.N. Life Devoted to Science. *Thermophys. Aeromechanics* **2014**, *21*, 265–278. [CrossRef]
11. Avdeev, A.A. Thermally Controlled Bubble Growth. In *Bubble Systems.*; Avdeev, A.A., Ed.; Springer International: Cham, Switzerland, 2016; Volume 4, pp. 99–132.
12. Khandekar, S.; Sahu, G.; Muralidhar, K.; Gatapova, E.Y.; Kabov, O.A.; Hu, R.; Luo, X.; Zhao, L. Cooling of High-Power LEDs by Liquid Sprays: Challenges and Prospects. *Appl. Therm. Eng.* **2021**, *184*, 115640. [CrossRef]
13. Wismer, K.L. The Pressure-Volume Relation of Super-Heated Liquids. *J. Phys. Chem.* **1922**, *26*, 301–315. [CrossRef]
14. Skripov, V.P. *Metastable Liquids*; Halsted Press: Sydney, Australia; John Wiley & Sons: New York, NY, USA, 1974.
15. Glod, S.; Poulikakos, D.; Zhao, Z.; Yadigaroglu, G. An Investigation of Microscale Explosive Vaporization of Water on an Ultrathin Pt Wire. *Int. J. Heat Mass Transf.* **2002**, *45*, 367–379. [CrossRef]
16. Asai, A. Bubble Dynamics in Boiling Under High Heat Flux Pulse Heating. *J. Heat Transf.* **1991**, *113*, 973–979. [CrossRef]
17. Iida, Y.; Okuyama, K.; Sakurai, K. Boiling Nucleation on a Very Small Film Heater Subjected to Extremely Rapid Heating. *Intern. J. Heat Mass Transf.* **1994**, *37*, 2771–2780. [CrossRef]
18. Nikitin, E.D. *Mechanisms of Vaporization in Superheated Water*; Ural Polytechnical Institute: Sverdlovsk, Russia, 1981.
19. Strenge, P.H.; Orell, A.; Westwater, J.W. Microscopic Study of Bubble Growth during Nucleate Boiling. *AIChE J.* **1961**, *7*, 578–583. [CrossRef]
20. Hong, Y.; Ashgriz, N.; Andrews, J. Experimental Study of Bubble Dynamics on a Micro Heater Induced by Pulse Heating. *J. Heat Transf.* **2004**, *126*, 259–271. [CrossRef]
21. Kuznetsov, V.V.; Oreshkin, V.I.; Zhigalin, A.S.; Kozulin, I.A.; Chaikovsky, S.A.; Rousskikh, A.G. Metastable States and Their Disintegration at Pulse Liquid Heating and Electrical Explosion of Conductors. *J. Eng. Thermophys.* **2011**, *20*, 240–248. [CrossRef]
22. Skripov, P.V.; Puchinskis, S.E.; Begishev, V.P.; Lipchak, A.I.; Pavlov, P.A. Heat Pulse Monitoring of Curing and Polymer–Gas Systems. *J. Appl. Polym. Sci.* **1994**, *51*, 1607–1619. [CrossRef]
23. Avedisian, C.T.; Cavicchi, R.E.; Tarlov, M.J. New Technique for Visualizing Microboiling Phenomena and Its Application to Water Pulse Heated by a Thin Metal Film. *Rev. Sci. Instr.* **2006**, *77*, 063706. [CrossRef]
24. Serdyukov, V.; Malakhov, I.; Surtaev, A. The Influence of Pressure on Local Heat Transfer Rate under the Vapor Bubbles during Pool Boiling. *Energies* **2023**, *16*, 3918. [CrossRef]
25. Skripov, P.V.; Bar-Kohany, T.; Antonov, D.V.; Strizhak, P.A.; Sazhin, S.S. Approximations for the Nucleation Temperature of Water. *Int. J. Heat Mass Transf.* **2023**, *207*, 123970. [CrossRef]
26. Kotov, A.N.; Gurashkin, A.L.; Starostin, A.A.; Skripov, P.V. Low-energy activation of superheated n-pentane boiling-up by laser pulse at the fiber-liquid interface. *Interf. Phen. Heat Transf.* **2022**, *10*, 15–23. [CrossRef]
27. Kotov, A.N.; Starostin, A.A.; Gorbatov, V.I.; Skripov, P.V. Thermo-Optical Measurements and Simulation in a Fibre-Optic Circuit Using an Extrinsic Fabry–Pérot Interferometer under Pulsed Laser Heating. *Axioms* **2023**, *12*, 568. [CrossRef]

28. Skripov, P.V.; Skripov, A.P. The Phenomenon of Superheat of Liquids: In Memory of Vladimir P. Skripov. *Int. J. Thermophys.* **2010**, *31*, 816–830. [CrossRef]
29. Gurashkin, A.L.; Starostin, A.A.; Uimin, A.A.; Yampol'skiy, A.D.; Ermakov, G.V.; Skripov, P.V. Experimental Determination of Superheated Liquid Density by the Optical Fiber Method. *J. Eng. Thermophys.* **2013**, *22*, 194–202. [CrossRef]
30. Lipnyagov, E.V.; Parshakova, M.A. Investigation of the Kinetics of Spontaneous Boiling-up of Superheated n-Pentane in a Glass Tube with Defects of the Inner Surface. I. Monitoring the Liquid-Vapor Surface Tension. *Int. J. Heat Mass Transf.* **2022**, *196*, 123254. [CrossRef]
31. Antonov, D.V.; Fedorenko, R.M.; Strizhak, P.A. Micro-Explosion Phenomenon: Conditions and Benefits. *Energies* **2022**, *15*, 7670. [CrossRef]
32. Davitt, K.; Arvengas, A.; Caupin, F. Water at the Cavitation Limit: Density of the Metastable Liquid and Size of the Critical Bubble. *Europhys. Lett.* **2010**, *90*, 16002. [CrossRef]
33. Gurashkin, A.L.; Yampol'skii, A.D.; Starostin, A.A.; Skripov, P.V. Optical Studies of the Initial Stage of Spontaneous Boiling-Up. *Tech. Phys. Lett.* **2013**, *39*, 751–754. [CrossRef]
34. Zubalic, E.; Vella, D.; Babnik, A.; Jezeršek, M. Interferometric Fiber Optic Probe for Measurements of Cavitation Bubble Expansion Velocity and Bubble Oscillation Time. *Sensors* **2023**, *23*, 771. [CrossRef]
35. McMillan, C.F.; Goosman, D.R.; Parker, N.L.; Steinmetz, L.L.; Chau, H.H.; Huen, T.; Whipkey, R.K.; Perry, S.J. Velocimetry of Fast Surfaces Using Fabry–Perot Interferometry. *Rev. Sci. Instr.* **1988**, *59*, 1–21. [CrossRef]
36. Surtaev, A.S.; Serdyukov, V.S.; Zhou, J.; Pavlenko, A.N.; Tumanov, V.V. An experimental study of vapor bubbles dynamics at water and ethanol pool boiling at low and high heat fluxes. *Int. J. Heat Mass Transf.* **2018**, *126*, 297–311. [CrossRef]
37. Vinogradov, V.E.; Pavlov, P.A. Rate of Bubble Growth at Limiting Superheats of a Stretched Liquid. *Heat Transf. Res.* **2007**, *38*, 389–398. [CrossRef]

energies

MDPI

Article

Modeling of Turbulent Heat-Transfer Augmentation in Gas-Droplet Non-Boiling Flow in Diverging and Converging Axisymmetric Ducts with Sudden Expansion

Maksim A. Pakhomov * and Viktor I. Terekhov

Laboratory of Thermal and Gas Dynamics, Kutateladze Institute of Thermophysics, Siberian Branch of Russian Academy of Sciences, Acad. Lavrent'ev Avenue 1, 630090 Novosibirsk, Russia
* Correspondence: pakhomov@ngs.ru

Abstract: The effect of positive (adverse) and negative (favorable) longitudinal pressure gradients on the structure and heat transfer of gas-droplet (air and water) flow in axisymmetric duct with sudden expansion are examined. The superimposed pressure gradient has a large influence on the flow structure and heat transfer in a two-phase mist flow in both a confuser and a diffuser. A narrowing of the confuser angle leads to significant suppression of flow turbulence (more than four times that of the gas-drop flow after sudden pipe expansion without a pressure gradient at $\varphi = 0°$). Recirculation zone length decreases significantly compared to the gas-droplet flow without a longitudinal pressure gradient (by up to 30%), and the locus of the heat-transfer maximum shifts slightly downstream, and roughly aligns with the reattachment point of the two-phase flow. Growth of the diffuser opening angle leads to additional production of kinetic energy of gas flow turbulence (almost twice as much as gas-droplet flow after a sudden pipe expansion at $\varphi = 0°$). The length of the flow recirculating region in the diffuser increases significantly compared to the separated gas-droplet flow without a pressure gradient ($\varphi = 0°$), and the location of maximum heat transfer shifts downstream in the diffuser.

Keywords: heat transfer; droplets evaporation; turbulence; droplet-laden flow; confuser; diffuser; pipe; sudden expansion; RANS

Citation: Pakhomov, M.A.; Terekhov, V.I. Modeling of Turbulent Heat-Transfer Augmentation in Gas-Droplet Non-Boiling Flow in Diverging and Converging Axisymmetric Ducts with Sudden Expansion. *Energies* **2022**, *15*, 5861. https://doi.org/10.3390/en15165861

Academic Editor: Gianpiero Colangelo

Received: 11 July 2022
Accepted: 9 August 2022
Published: 12 August 2022

1. Introduction

Two-phase flows in pipes or channels with a backward-facing step (BFS) are often used in energy and chemical equipment. They have a rather simple flow geometry and are one of the classical types of shear flows, but their flow structure is quite complex. A flow detaches from the sharp edge at the flow SE station, thus forming a region of shear mixing layer. A large recirculation flow region (a few step heights) develops (see comprehensive reviews [1,2]).

The complexity of modeling flow and heat transfer is exacerbated after BFS in the presence of a longitudinal pressure gradient (LPG) in an expanding (diffuser) or narrowing (confuser) subsonic turbulent two-phase flow (see Figure 1). An overview of the state of research on flows in a diffuser or confuser without sudden expansion of a pipe [3] or a channel [4,5] has been presented. The study of the effect of LPG behind a pipe or channel with SE on mean and fluctuational flow and heat transfer is an important for mechanical engineering. There are several studies on the development of separated flows with the influence of longitudinal pressure gradient for a single-phase flow, yet only a few of these experimental works concerned the flow in diffusers and confusers with a BFS [6–9].

An effect of flow separation in the field of LPG was experimentally evaluated in said studies. The position of an "upper" duct wall was changed, which caused narrowing or expansion of a cross-section, whereas the "lower" wall with the SE remained unchanged. The most detailed structures of the turbulent flow were assessed in [6] using the LDA method along the length of the diffuser channel. The authors measured the profiles of averaged longitudinal and transverse velocities and their fluctuations, Reynolds stresses, length of the recirculation region, triple correlations, and turbulent viscosity. The authors then compared their experimental and numerical data.

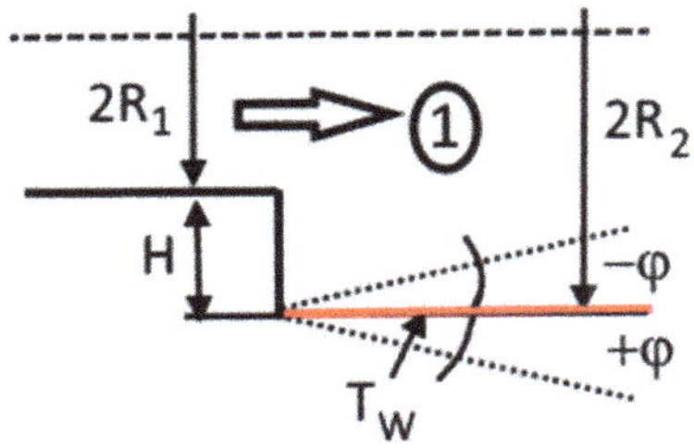

Figure 1. Schematic view of the flow in diffuser (APG, $+\varphi$), confuser (FPG, $-\varphi$), and in the separated flow in pipe sudden expansion (ZPG, $\varphi = 0$). 1 is the droplet-laden flow.

The experimental results on the effect of favorable pressure gradient (FPG) and adverse pressure gradient (APG) in a channel behind a BFS on heat transfer and wall pressure distributions at Reynolds numbers $Re_H = U_{m1}H/\nu = (0.4–1.2) \times 10^4$ were presented in [10]. The diffuser opening angle varied in the range of $\varphi = 0–4°$, and confuser narrowing angle was varied within $\varphi = 0––7.5°$. The magnitude of the Nusselt number increases as the LPG increases for a narrowing channel, and it decreases for the diffuser. The locus of the heat-transfer maximum moves downstream with diffuser expansion, and shifts upstream towards the step as the confuser narrows. In [11], a quantitative study assessed the effect of an APG on mean flow, turbulence, and heat transfer in an axisymmetric diffuser in a pipe with SE. The literature also presents experimental [12] and numerical [13–18] studies of fluid flow and heat transfer in single-phase turbulent flows without SE of a pipe or channel in the presence of APG and FPG for a single-phase flow.

Solid particles addition to a turbulent flow in a BFS have large effect on reduction of turbulent kinetic energy (TKE) in backward-facing step flow [19]. Droplets evaporation in turbulent flow behind a BFS [20] or after a pipe with SE [21] causes significant intensification of heat transfer (by several times in comparison with a single-phase flow). Authors of this work have published numerical investigations of heat-transfer augmentation in gas-droplet flows behind a pipe with SE [21]. There are few papers concerning numerical simulation of gas-liquid flow [22,23] and droplet-laden [24] flows in a converge or divergent channel without sudden expansion; we know of only one work on the numerical study of heat transfer in two-phase flows after pipe sudden expansion with LPG [25], where the effect of evaporation of water droplets on heat transfer in an axisymmetric diffuser was studied. Heat transfer in turbulent droplet-laden flow with SE with APG and FPG has not been previously performed. The influence of LPG on flow and heat transfer in the confuser and diffuser after pipe SE is evaluated in the present study.

2. Mathematical Methods and Numerical Solution

The motion and heat transfer of a two-phase turbulent gas-droplet flow in a pipe with SE is numerically considered. A sketch of the flow is given in Figure 1. To simulate the dispersed phase dynamics, the Eulerian approach [26–28] is used. The Eulerian approach is widely used for the simulations of two-phase confined flows [21,22,29,30]. The system of axisymmetric stationary Reynolds-averaged Navier–Stokes (RANS) equations accounts for the effect of vaporizing drops on mean and fluctuational transport processes [21,25]. The set of governing equations both for gas and dispersed phases have been provided in detail [21,25]. The volume fraction of the droplets is low ($\Phi_1 = M_{L1}\rho/\rho_L < 1.2 \times 10^{-4}$ for the highest mass fraction studied M_{L1} = 10%). Drops are rather small (d_1 < 100 μm), so effects of their collisions can be neglected. Gas phase turbulence is predicted using the elliptical Reynolds stress model [31] by taking the dispersed phase influence on TKE [32].

Break-up and coalescence of droplets in flow is not taken into account due to their rarity ($\Phi_1 < 1.2 \times 10^{-4}$) [33]. The Weber number $\mathrm{We} = \rho(\mathbf{U}_S - \mathbf{U}_L)^2 d/\sigma << 1$ and the bag break-up are ignored [33,34]. Here, $\mathbf{U}_S = \mathbf{U} + \langle u'_S \rangle$ is the gas velocity seen by the droplet, and $\langle u'_S \rangle$ is the drift velocity between fluid flow and drops [35]. This assumption is applicable when the pipe cross-section expands for a diffuser. The use of this approach seems less obvious for a confuser, even when taking into account the preliminary pipe with SE. Effect of break-up and coalescence in the flow can be neglected due to a low droplet volume fraction at the inlet according to preliminary author's estimations.

The technique for numerical implementation of the Eulerian approach for two phases is described in detail in [21,25]. The numerical solution was obtained using the finite volume method on staggered grids. The QUICK scheme of third-order ode accuracy was utilized for solution of convective terms. Central differences of second-order accuracy were evaluated for diffusion fluxes. Pressure–velocity fields were corrected according to SIMPLEC procedure.

All simulations were carried out on a "basic" mesh containing 550 × 200 control volumes (CV) for the diffuser with the largest opening angle, and for the confuser with the largest convergence angle of 550 × 100. The information about meshes for the confuser and diffuser is summarized in Table 1. The difference in calculations of the Nusselt number for the two-phase gas-droplet flow did not exceed 0.1%. A further increase in their number does not significantly affect the results of numerical calculations. The grid verification for the case of droplet-laden flow in pipe with SE was presented in [21]. The grid independence tests for two-phase flows in APG and FPG are given in Figure 2 for the smallest constriction angle in the confuser and for the largest opening angle in diffuser. The Nusselt number at a constant wall temperature is determined by dependence:

$$\mathrm{Nu} = -(\partial T/\partial y)_W H/(T_W - T_m),$$

where T_m is mass-averaged temperature of gas in the considered cross-section.

Table 1. Meshes for two-phase flow in the confuser (FPG) and diffuser (APG).

Flow Type	"Basic"	"Coarse"	"Fine"
Confuser	550 × 100	300 × 50	850 × 150
Diffuser	550 × 200	300 × 100	850 × 300

The convergence criteria for all residual levels in this study were up to 10^{-5}. The differences in Nusselt number and gas-phase kinetic energy of turbulence for gas-droplet APG and FPG flows were up to 10^{-6}.

The model was validated against experimental results on the flow and heat transfer for the single-phase axisymmetric diffuser downstream of a pipe with SE. The difference between our predictions and measured results of previous experiments did not exceed 15%. These results were given in our previous paper [25], but this comparison is not presented here. We did not find measured or numerical results concerning the study of an APG or FPG of gas-droplet flow in a pipe or duct with sudden expansion. We performed the comparisons with experimental two-phase droplet-laden mist and solid particle-laden turbulent flow behind the BFS and a pipe with SE. These results were published in a previous paper [21] but are not included here. We believe that the validation analysis of two-phase solid particle-laden and droplet-laden flows behind backward-facing step or pipe sudden expansion without LPG have been fully completed.

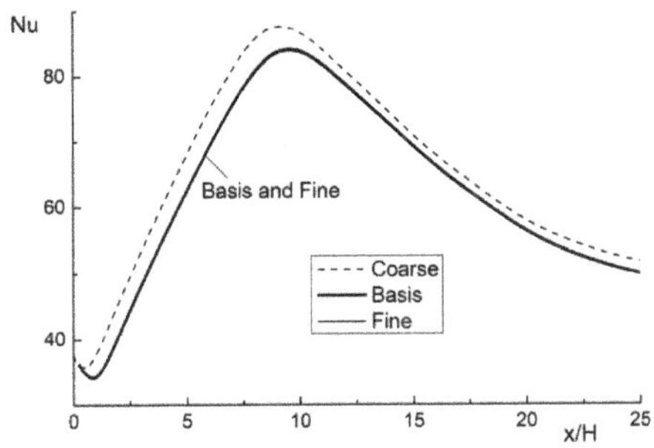

(a)

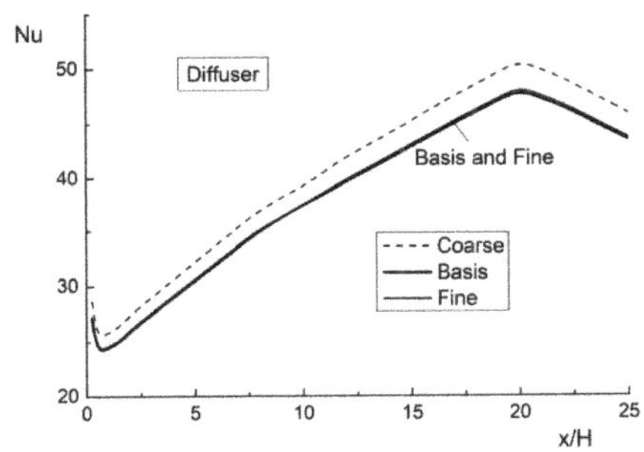

(b)

Figure 2. Grid independence tests for confuser at $\varphi = -1°$ (**a**) and diffuser at $\varphi = 5°$ (**b**).

3. Results and Discussion

The primary concern of this study was shown the effect of diffuser opening and confuser narrowing angles on the characteristics and heat transfer in the two-phase mist with vaporized water droplets after the pipe with SE. Drop diameter and mass fraction decreased due to evaporation both in the axial and radial directions after the flow detachment section.

The diffuser expanding angle was φ = 0–5° and the confuser convergence angle φ = 0–−3°. The pipe diameter before SE was $2R_1$ = 20 mm, after SE it was $2R_2$ = 60 mm, and the step height was H = 20 mm. The computational domain after pipe expansion was $25H$ = 0.5 m. Mass-average air velocity before separation was U_{m1} = 15 m/s, and the Reynolds number was $\mathrm{Re}_H = HU_{m1}/\nu \approx 2 \times 10^4$. The wall temperature was T_W = const = 373 K, and the temperatures of air and droplets at the inlet were $T_1 = T_{L1}$ = 293 K. Water droplets were added to a single-phase air turbulent flow at the inlet, and their initial velocity was set constant over pipe cross-section: $U_{L1} = 0.8U_{m1}$. Inlet droplet size was constant d_1 = 1–100 μm, and mass fraction M_{L1} = 0.01–0.1. The Stokes number in mean motion was $\mathrm{Stk} = \tau/\tau_f$ = 0.03–3, where $\tau_f = 5H/U_{m1}$ is the turbulent time macroscale [19,20]. Here, $\tau = \rho_L d^2/(18\rho\nu W)$ is the particle relaxation time, $W = 1 + \mathrm{Re}_L^{2/3}/6$ and $\mathrm{Re}_L = |\mathbf{U}_S - \mathbf{U}_L| d/\nu$ is the dispersed-phase Reynolds number. The Stokes number $\mathrm{Stk}_K = \tau/_{fK}$ = 0.2–20, where τ_K is the Kolmogorov timescale. While the value of interfacial velocity in our previous works [21,25] was based only on the average velocity of the carrier phase, it is based on the actual value in the present study.

3.1. The Wall Friction and Pressure Coefficients

The distributions of wall friction coefficient $C_f/2 = \tau_W/(\rho U_{m1}^2)$ and pressure coefficient $C_P = 2(P_W - P_1)/(\rho U_{m1}^2)$ along the length of diffuser and confuser in gas-droplet flow with variation of expansion and contraction angles are shown in Figure 3. Here, τ_W is the wall friction; P_W, P_1 are the mean static pressures on the wall in considered and inlet cross-sections. The data for the flow after a pipe with SE, without an effect of LPG ($\varphi = 0°$), are also shown in this figure for comparison.

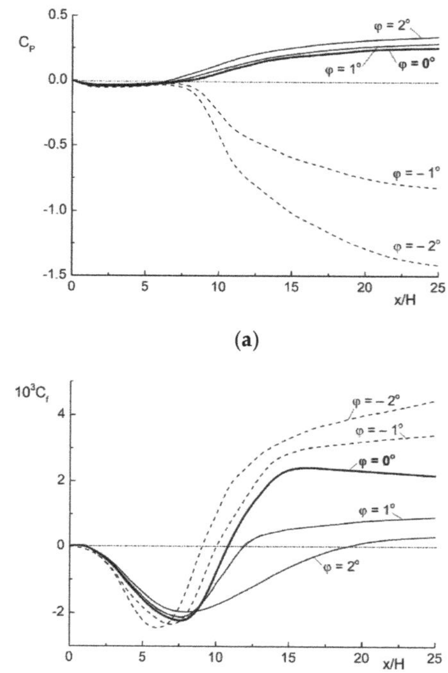

(a)

(b)

Figure 3. The evolution of pressure C_P (**a**) and wall friction C_f (**b**) coefficients along the axial coordinates in ZPG ($\varphi = 0°$), diffuser ($\varphi > 0°$, APG), and confuser ($\varphi < 0°$, FPG). $M_{L1} = 0.05$, $d_1 = 30$ μm.

The distributions of non-dimensional pressure coefficients along axial coordinates with the development of a separated flow in the diffuser and confuser are shown in Figure 3a. In the diffuser, directly behind the flow separation point, a negative pressure region is formed, with a length of $x/H \approx 7$. The pressure coefficient also increases with an increase in the diffuser opening angle, which can be attributed mainly to flow deceleration. The zone with pressure attenuation is formed in the confuser directly behind the flow detachment cross-section, and its length axial direction is x/H = 5–7. With growth of the confuser convergence angle, the presence of a significant region of pressure attenuation is observed, and the absolute value of pressure attenuation increases noticeably as the convergence angle increases (more than 5.5 times at $\varphi = -2°$). Obviously, the main reason for a significant pressure decrease in confuser is flow acceleration. The wall friction coefficient C_f decreases significantly (several times over) with growth of APG, and a sharp increase in the flow recirculation zone is observed (see Figure 3b). With the increase in the magnitude of FPG, the wall friction coefficient increases noticeably (almost doubling) after the zone of flow relaxation.

3.2. *The Flow Structure in Confuser (FPG) and Diffuser (APG)*

Profiles of the mean axial velocity, temperature, and turbulent kinetic energy of the gas phase in a cross-section at x/H = 15 are shown in Figure 4. The predictions are carried out for different values for the diffuser (APG, $\varphi > 0°$), the confuser (FPG, $\varphi < 0°$), and in the separated flow behind sudden expansion of the pipe (φ = 0°). A large effect of two-phase flow detachment with a zero pressure gradient (ZPG) and with FPG and APG on the mean axial velocity distributions is revealed in two-phase flow. Obviously, the increase in the diffuser opening angle leads to a reduction of gas velocity in the core zone (see Figure 4a). It should be noted that in a cylindrical duct, as well as at small diffuser opening angles ($\varphi \leq 2°$), the separated flow is reattached in cross-section (x/H = 15) and the flow is relaxed. Air velocity and the velocity gradient in the radial direction in the core region increase in the confuser.

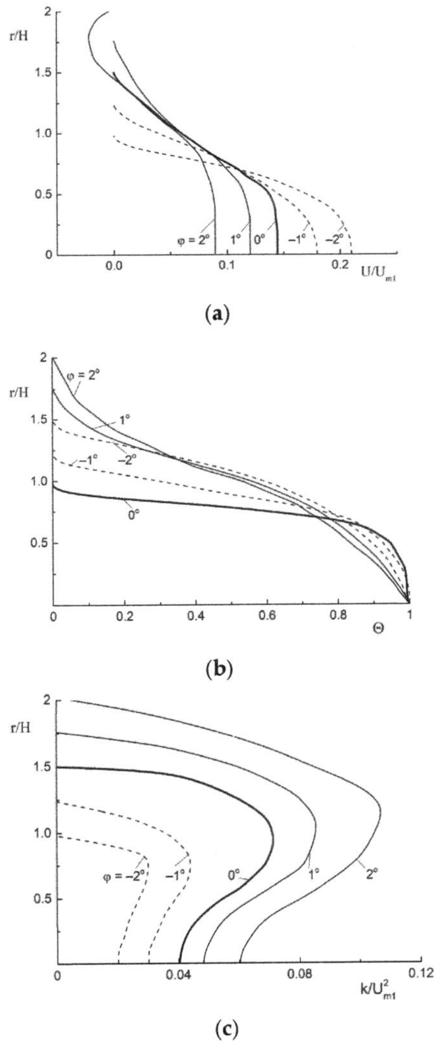

Figure 4. Mean streamwise velocity component (**a**), temperature (**b**), TKE (**c**) of the gas phase in ZPG (φ = 0°), confuser (FPG, φ < 0°), and diffuser (APG, φ > 0°). The results for the confuser are the dashed lines, for the diffuser are the solid curves, and the separated flow with φ = 0° are the bolded lines. M_{L1} = 0.05, d_1 = 30 μm.

Gas temperature distributions $\Theta = (T - T_W)/(T_0 - T_W)$ over the pipe radius depend, to a lesser extent, on the longitudinal pressure gradient rather than on distributions of the axial gas velocity (see Figure 4b). Here, T_0 and T_W are gas phase temperatures on a pipe axis and on a wall. The slightly changing diverging angle of the diffuser ($\varphi \leq 2°$) and the converging angle of the confuser ($\varphi \geq -2^0$) have little effect on the gas phase temperature in droplet-laden flow. The temperature increases for the diffuser and decreases for the confuser. This leads to heat-transfer enhancement in the confuser and heat-transfer suppression in the diffuser. These conclusions qualitatively concur with the results of simulations [11] for a single-phase flow in a diffuser behind a pipe with SE. Gas temperature becomes lesser for gas-droplet flow compared to the case at $\varphi = 0°$.

Turbulent kinetic energy (TKE) of the gaseous phase is significantly enhanced (by up to two times over) by an increase in the diffuser opening angle (see Figure 4c). The TKE of the gas phase is calculated for an axisymmetric flow using a known formula: $2k = \langle u'^2 \rangle + \langle v'^2 \rangle + \langle w'^2 \rangle \approx \langle u'^2 \rangle + 2\langle v'^2 \rangle$. This is not an effect of the dispersed phase; it is known that the presence of a finely dispersed phase suppresses the carrier-phase turbulence in the separated flow, both behind the BFS [19,20] and with sudden expansion of the pipe [21,22]. Particles or droplets are involved in the mean gas movement and a part of the turbulent energy of a carrier flow is spent on this process [19,32]. The maximum kinetic energy of turbulence is observed in the mixing layer, and the same phenomena were found for the gas-droplet flow in the pipe with sudden expansion at ZPG ($\varphi = 0°$) [21]. This effect was shown previously, in our recent study of an axisymmetric diffuser with a sudden pipe expansion [25]. An increase in the LPG in an axisymmetric diffuser with pipe SE causes additional flow turbulization. The maximum value of the turbulent kinetic energy of the carrier phase for a confuser decreases almost twice as much compared to the turbulence level of a separated two-phase flow at ZPG. Such a significant TKE suppression of the carrier phase cannot be explained only by the effect of the dispersed phase.

The transverse distributions of mean axial water droplet velocity U_L/U_{L1} (a), drops in temperature $\Theta_L = (T_L - T_{L,\max})/(T_{L,0} - T_{L,\max})$ (b), and the mass fraction M_L/M_{L1} (c) of dispersed phase in the confuser and diffuser in a pipe with SE are presented in Figure 5. Here T_L, T_L, and $T_{L,\max}$ are the droplet temperature, the droplet temperature on pipe axis, and maximum droplet temperature in corresponding cross-section, respectively.

With growth of the confuser convergence angle, a significant increase in the longitudinal averaged velocity of droplets occurs (by more than double at $\varphi = -2°$ as compared to the separated flow at ZPG) (see Figure 5a). The droplet temperature profile has a qualitatively similar form for all three types of ducts (ZPG, APG, and FPG) studied previously (see Figure 5b). On the whole, droplet temperature distributions are similar to those for the gas phase. The maximum value of droplet mass fraction is obtained in the axial region of the pipe, and the minimum value is obtained in its near-wall region (see Figure 5c). The simulations for droplets' mass fractions $M_{L1} > 10\%$ were not successful due to the possible effect of droplets deposition in reality. Most likely, the distribution of the mass fraction of droplets is qualitatively similar to those for $M_{L1} = 10\%$, but there are quantitative differences. It is also necessary to take into account the effect of droplet deposition on the wall from a two-phase flow, and the possible formation of liquid spots and films on the wall surface. The influence of droplet deposition on transport processes and heat transfer are not taken into account for our numerical results obtained for $M_{L1} = 5\%$. Obviously, for high values of the droplets' mass fraction at the inlet, it is necessary to account for the influence of the deposition process and the entrainment of liquid droplets into the droplet-laden flow from the liquid film or spots.

3.3. The Effect of LPG on the Mean Parametrs of the Two-Phase Mist Flow

A significant increase in the length of recirculating area x_R is observed in two-phase flow in diffuser (see Figure 6). The locus of the heat-transfer peak $x_{\max}$ moves in the downstream direction. A slight increase in the flow recirculation region is shown for small expanding angles ($\varphi \leq 1°$), and the position of the heat-transfer maximum is close to the locus of the reattachment point of two-phase flow. The coordinate of $x_{\max}$ moves downstream by almost double ($\varphi = 5°$) in comparison with the case of $\varphi = 0°$. The presence of FPG leads to a reduction in flow recirculation area and the coordinates of x_{max} move upstream by about 30–35% compared to the case of $\varphi = 0°$. The significant displacement of flow reattachment points in the diffuser and confuser is caused by deformation of the gas phase velocity profile due to the effect of LPG.

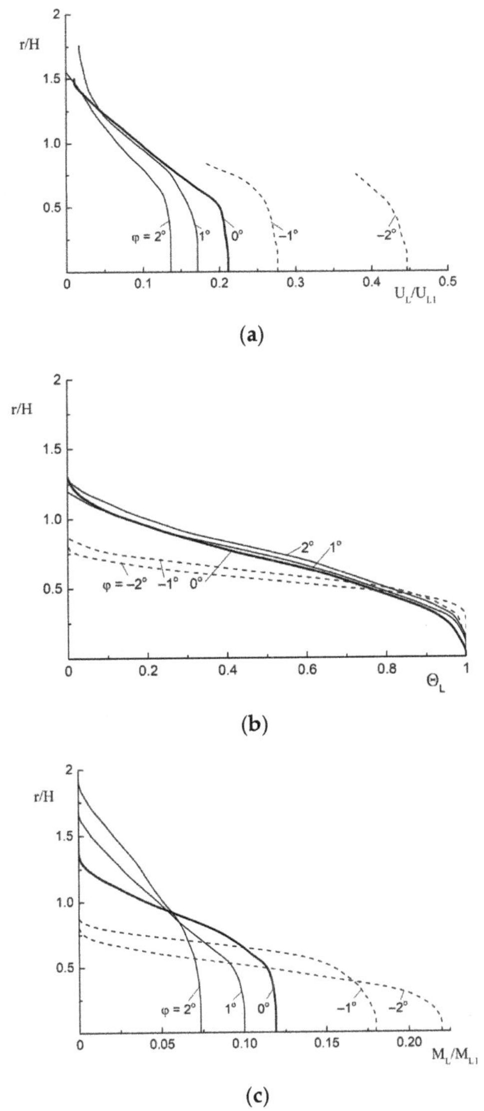

Figure 5. Mean axial velocity componet (**a**), temperature (**b**), and mass fraction (**c**) of the dispersed phase. The results for confuser (FPG, $\varphi < 0°$) are the dashed lines, for the diffuser (APG, $\varphi > 0^0$) are the solid curves, and the separated flow with $\varphi = 0°$ are the bold lines. $M_{L1} = 0.05$, $d_1 = 30$ μm.

The TKE of the carrier phase increases almost two times over in the diffuser at $\varphi = 5°$ as compared with the case without longitudinal pressure gradient $\varphi = 0°$. Changing the confuser convergence angle causes suppression of the level of turbulence more than three times over. The heat transfer decreases significantly with expanding of diffuser opening angle (almost by a factor of 1.5 as compared to the separated flow in the pipe at $\varphi = 0°$).

For the confuser, an increase in the relative value of the maximum heat transfer is observed at $\varphi = -3°$ (by approximately 20%). The heat transfer for the confuser (FPG) case has the greatest value, and the diffuser (APG) has the smallest.

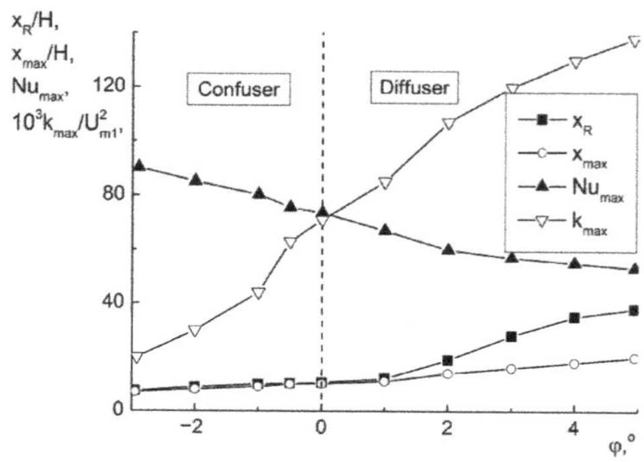

Figure 6. Effect of LPG on recirculating length x_R, location of heat transfer maximum x_{max}, value of maximal Nusselt number Nu_{max}, and maximum of TKE k_{max} in gas-droplet flow in pipe SE. $M_{L1} = 0.05$, $d_1 = 30$ μm.

The effect convergence (confuser) and divergence (diffuser) on Nusselt number distributions along the axial coordinate for the separated flow (ZPG), confuser (FPG), and diffuser (APG) are shown in Figure 7. Initially, for two-phase mist flows with APG and FPG, the attenuation of heat transfer rate is observed. This is typical both for both types of flows and for the case of gas-droplet flow in pie with SE at $\varphi = 0°$. Then there is a sharp increase in heat transfer with the achievement of a maximum heat transfer. In the zone of flow relaxation, the observed decrease in Nusselt number is similar to a single-phase flow.

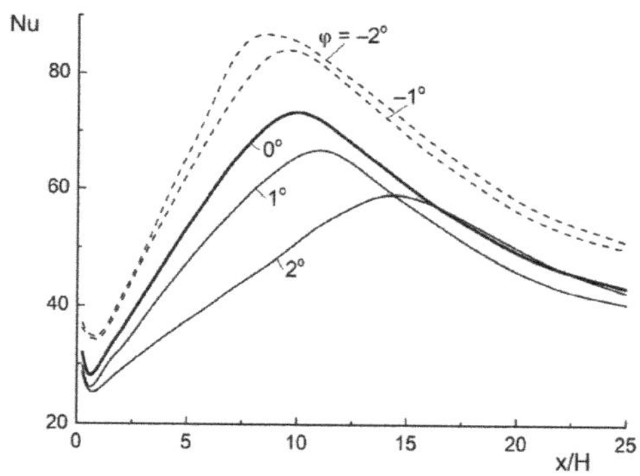

Figure 7. Nusslet numbers distributions along streamwise coordinate in ZPG ($\varphi = 0°$), FPG ($\varphi < 0°$), and APG ($\varphi > 0°$). $M_{L1} = 0.05$, $d_1 = 30$ μm.

The influence of water droplets' mass concentration on the maximal magnitude of heat transfer Nu_{max} for a diffuser and confuser after pipe SE is presented in Figure 8. For all types of flow behind the pipe with SE at $\varphi = 0°$, in the diffuser ($\varphi = 2°$), and in the confuser ($\varphi = -2°$), an increase in the maximum heat-transfer value (up to 75% in a single-phase airflow) was obtained with increasing mass fraction of droplets to $M_{L1} = 10\%$. The heat transfer in confuser enhances compare to the diffuser and for two-phase separated flow with ZPG at $\varphi = 0°$.

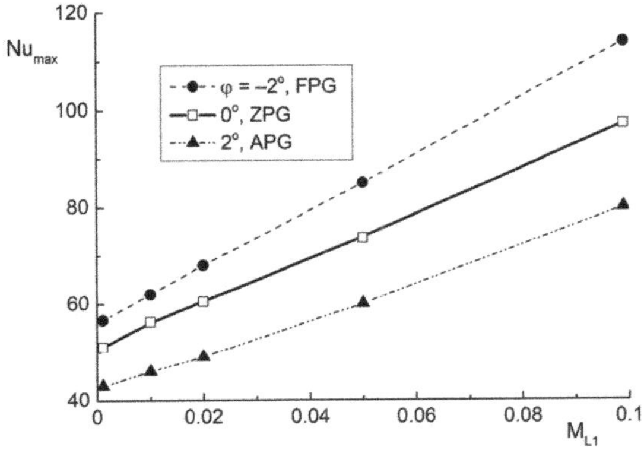

Figure 8. The magnitude of maximal heat transfer in the diffuser ($\varphi = 2°$), confuser ($\varphi = -2°$) and separated flow ($\varphi = 0°$) vs water droplets mass concentration. $d_1 = 30$ μm.

4. Conclusions

The numerical results of the effects of favorable and adverse longitudinal pressure gradients on the flow and heat transfer augmentation, in a droplet-laden flow in a pipe with SE, are presented. Elliptical second-moment closure was used to predict the gas phase turbulence with taking into account the effect of droplets presence. While this study does not have a direct application, it shows potential ways to control the turbulence level and to enhance heat transfer performance in APG and FPG flow behind a backward-facing step. Thus, these data may be of interest for various practical applications. The scope of the model's use is limited the inlet droplet diameter $d_1 = 100$ μm and their initial mass fraction $M_{L1} \leq 10\%$. This can be explained by noting that the model does not take into account the formation and evolution of a liquid film on the pipe wall, as drops break up and coalesce.

The presence of flow expansion (diffuser) and constriction (confuser) of pipe with SE shows significant effect on the mean and fluctuational flow characteristics, and heat transfer in an axisymmetric. The increase of the confuser constriction angle causes considerable reduction of the pressure coefficient. The length of the flow recirculating area noticeably shortens compared to the gas-droplet flow behind the pipe with SE at angle $\varphi = 0°$, and the point of maximum of heat transfer slightly shifts downstream. The heat transfer augmentation and the suppression of turbulence in a two-phase flow in a confuser are mainly due to the FPG. The large growth of flow recirculating area (up to 3.5 times at $\varphi = 5°$) compared to the gas-droplet flow downstream of pipe SE at $\varphi = 0°$ is obtained. The expansion of the diffuser leads to reduction of the wall friction coefficient. Two-phase flow does not reattach to the wall at angle $\varphi = 5°$. Points of flow reattachment and maximum heat transfer are significantly shifted downstream by an increase in the opening angle of the diffuser. The significant heat-transfer suppression (by up to 1.5 times) and turbulence production (by up to two times) are observed for the two-phase mist flow in a diffuser.

Author Contributions: Conceptualization, M.A.P. and V.I.T.; methodology, M.A.P. and V.I.T.; Investigation, M.A.P.; data curation, M.A.P. and V.I.T.; formal analysis, M.A.P. and V.I.T.; writing—original draft preparation, M.A.P. and V.I.T.; writing—review and editing, M.A.P. and V.I.T.; resources, M.A.P. and V.I.T.; project administration, V.I.T. All authors have read and agreed to the published version of the manuscript.

Funding: This work was financially supported by the grant of the Russian Science Foundation (project code 21-19-00162).

Institutional Review Board Statement: Not applicable.

Informed Consent Statement: Not applicable.

Conflicts of Interest: The authors declare no conflict of interest.

Nomenclature

d	droplet diameter
H	step height
M_L	mass fraction
$\mathrm{Nu} = -(\partial T/\partial y)_W H/(T_W - T_m)$	Nusselt number
$\mathrm{Re}_H = U_{m1}H/\nu$	Reynolds number
$\mathrm{Stk} = \tau/\tau_f$	mean Stokes number
T	temperature
$\mathbf{U}$	average velocity vector
U_i, U_j	mean gas velocities components
$\mathbf{U}_S = \mathbf{U} + \langle u'_S \rangle$	gas velocity vector seen by the droplet
$\langle u'_S \rangle$	drift velocity between fluid flow and drops
$\mathrm{We} = \rho(\mathbf{U}_S - \mathbf{U}_L)^2/\sigma$	Weber number
x	streamwise coordinate
$x_{\max}$	location of heat-transfer maximum
x_R	reattachment length
Subscripts	
0	single-phase fluid (air) flow
1	initial condition
L	liquid
m	mean
max	maximal value
W	wall
Greek	
Φ	volume fraction
λ	thermal conductivity
ρ	density
ν	kinematic viscosity
τ	particle relaxation time
τ_W	wall shear stress
φ	diffuser opening angle ($\varphi > 0$) or confuser ($\varphi < 0$) narrowing angle
Acronym	
APG	adverse pressure gradient
BFS	backward-facing step
FPG	favorable pressure gradient
LPG	longitudinal pressure gradient
CV	control volume
SE	sudden expansion
ZPG	zero pressure gradient

References

1. Eaton, J.K.; Johnston, J.P. A review of research on subsonic turbulent flow reattachment. *AIAA J.* **1981**, *19*, 1093–1100. [CrossRef]
2. Chen, L.; Asai, K.; Nonomura, T.; Xi, G.N.; Liu, T.S. A review of backward-facing step (BFS) flow mechanisms, heat transfer and control. *Thermal Sci. Eng. Progr.* **2018**, *6*, 194–216. [CrossRef]
3. Klein, A. Review: Effects of inlet conditions on conical-diffuser performance. *ASME J. Fluids Eng.* **1981**, *103*, 250–257. [CrossRef]
4. Azad, R.S. Turbulent flow in a conical diffuser: A review. *Exp. Therm. Fluid Sci.* **1996**, *13*, 318–337. [CrossRef]
5. Apsley, D.D.; Leschziner, M.A. Advanced turbulence modelling of separated flow in a diffuser. *Flow Turbul. Combust.* **1999**, *63*, 81–112. [CrossRef]
6. Driver, D.M.; Seegmiller, H.L. Features of a reattaching turbulent shear layer in divergent channel flow. *AIAA J.* **1985**, *23*, 163–171. [CrossRef]
7. Ra, S.H.; Chang, P.K. Effects of pressure gradient on reattaching flow downstream of a rearward-facing step. *J. Aircr.* **1990**, *27*, 93–95. [CrossRef]
8. Iftekhar, H.; Agelin-Chaab, M. Structure of turbulent flows over forward facing steps with adverse pressure gradient. *ASME J. Fluids Eng.* **2016**, *138*, 111202. [CrossRef]
9. Wang, L.B.; Tao, W.Q.; Wang, Q.W.; Wong, T.T. Experimental study of developing turbulent flow and heat transfer in ribbed convergent/divergent square ducts. *Int. J. Heat Fluid Flow* **2001**, *22*, 603–613. [CrossRef]

10. Terekhov, V.; Dyachenko, A.; Smulsky, Y. The effect of longitudinal pressure gradient on heat transfer in a separated flow behind a sudden expansion of the channel. *Heat Transf. Eng.* **2020**, *41*, 1404–1416.
11. Terekhov, V.I.; Bogatko, T.V. Aerodynamics and heat transfer in a separated flow in an axisymmetric diffuser with sudden expansion. *J. Appl. Mech. Techn. Phys.* **2015**, *56*, 471–478. [CrossRef]
12. Güemes, A.; Sanmiguel Vila, C.; Örlü, R.; Vinuesa, R.; Schlatter, P.; Ianiro, A.; Discetti, S. Flow organization in the wake of a rib in a turbulent boundary layer with pressure gradient. *Exp. Therm. Fluid Sci.* **2019**, *108*, 115–124. [CrossRef]
13. Leont'ev, A.I.; Lushchik, V.G.; Reshmin, A.I. Heat transfer in conical expanding channels. *High Temp.* **2016**, *54*, 270–276. [CrossRef]
14. Lushchik, V.G.; Reshmin, A.I. Heat transfer enhancement in a plane separation-free diffuser. *High Temp.* **2018**, *56*, 569–575. [CrossRef]
15. Lushchik, V.G.; Makarova, M.S.; Reshmin, A.I. Laminarization of flow with heat transfer in a plane channel with a confuser. *Fluid Dyn.* **2019**, *54*, 67–76. [CrossRef]
16. Sakhnov, A.Y. Local laminarization within the mild pressure gradient flow over the heated wall. *Int. J. Heat Mass Transf.* **2021**, *165*, 120698. [CrossRef]
17. Sakhnov, A.Y.; Naumkin, V.S. Velocity overshoot within the accelerated subsonic boundary layer over the heated wall. *Int. J. Heat Mass Transf.* **2020**, *161*, 120249. [CrossRef]
18. Hajaali, A.; Stoesser, T. Flow separation dynamics in three-dimensional asymmetric diffusers. *Flow Turbul. Combust.* **2022**, *108*, 973–999. [CrossRef]
19. Fessler, J.R.; Eaton, J.K. Turbulence modification by particles in a backward-facing step flow. *J. Fluid Mech.* **1999**, *314*, 97–117. [CrossRef]
20. Hishida, K.; Nagayasu, T.; Maeda, M. Augmentation of convective heat transfer by an effective utilization of droplet inertia. *Int. J. Heat Mass Transf.* **1995**, *38*, 1773–1785. [CrossRef]
21. Pakhomov, M.A.; Terekhov, V.I. Second moment closure modelling of flow, turbulence and heat transfer in droplet-laden mist flow in a ver pipe with sudden expansion. *Int. J. Heat Mass Transf.* **2013**, *66*, 210–222. [CrossRef]
22. Ahmadpour, A.; Noori Rahim Abadi, S.M.A.; Kouhikamali, R. Numerical simulation of two-phase gas–liquid flow through gradual expansions/contractions. *Int. J. Multiph. Flow* **2016**, *79*, 31–49. [CrossRef]
23. Kopparthy, S.; Mansour, M.; Janiga, G.; Thévenin, D. Numerical investigations of turbulent single-phase and two-phase flows in a diffuser. *Int. J. Multiph. Flow* **2020**, *130*, 103333. [CrossRef]
24. Golubkina, I.V.; Osiptsov, A.N. Compressible gas-droplet flow and heat transfer behind a condensation shock in an expanding channel. *Int. J. Therm. Sci.* **2022**, *179*, 107576. [CrossRef]
25. Pakhomov, M.A.; Terekhov, V.I. Gas-droplet flow structure and heat transfer in an axisymmetric diffuser with a sudden expansion. *J. Appl. Mech. Techn. Phys.* **2020**, *61*, 787–797. [CrossRef]
26. Drew, D.A. Mathematical modeling of two-phase flow. *Ann. Rev. Fluid Mech.* **1983**, *15*, 261–291. [CrossRef]
27. Reeks, M.W. On a kinetic equation for the transport of particles in turbulent flows. *Phys. Fluids A* **1991**, *3*, 446–456. [CrossRef]
28. Derevich, I.V.; Zaichik, L.I. Particle deposition from a turbulent flow. *Fluid Dyn.* **1988**, *23*, 722–729. [CrossRef]
29. Wu, H.; Yang, C.; Zhang, Z.; Zhang, Q. Simulation of two-phase flow and syngas generation in biomass gasifier based on two-fluid model. *Energies* **2022**, *15*, 4800. [CrossRef]
30. Zeng, Y.; Xu, W. Investigation on bubble diameter distribution in upward flow by the two-fluid and multi-fluid models. *Energies* **2021**, *14*, 5776. [CrossRef]
31. Fadai-Ghotbi, A.; Manceau, R.; Boree, J. Revisiting URANS computations of the backward-facing step flow using second moment closures. Influence of the numerics. *Flow Turbul. Combust.* **2008**, *81*, 395–410. [CrossRef]
32. Elgobashi, S. On predicting particle-laden turbulent flows. *Appl. Scient. Res.* **1994**, *52*, 309–329. [CrossRef]
33. Lin, S.P.; Reitz, R.D. Drop and spray formation from a liquid jet. *Ann. Rev. Fluid Mech.* **1998**, *30*, 85–105. [CrossRef]
34. Elgobashi, S. An updated classification map of particle-laden turbulent flows. In *IUTAM Symposium on Computational Approaches to Multiphase Flow, Fluid Mechanics and Its Applications*; Balachandar, S., Prosperetti, A., Eds., Springer: Dordrecht, The Netherlands, 2006; Volume 81, pp. 3–10.
35. Mukin, R.V.; Zaichik, L.I. Nonlinear algebraic Reynolds stress model for two-phase turbulent flows laden with small heavy particles. *Int. J. Heat Fluid Flow* **2012**, *33*, 81–91. [CrossRef]

MDPI
St. Alban-Anlage 66
4052 Basel
Switzerland
www.mdpi.com

Energies Editorial Office
E-mail: energies@mdpi.com
www.mdpi.com/journal/energies

www.ingramcontent.com/pod-product-compliance
Lightning Source LLC
Chambersburg PA
CBHW061626210926
51850CB00009B/2453
9783725807178